A Pinch of Salt:
Harm Reduction in
the Pandemic

A Pinch of Salt: Harm Reduction in the Pandemic

Written by Austin Mardon, Samira Sunderji,
Hannah Cho, Khushi Rathod, Neha Saroya, and
Catherine Mardon

Typeset and Design by Ethan Gabriel Saldana

GM PRESS

2020

A Golden Meteorite Press Book
Printed in Canada

First Printing: 2020

ISBN 978-1-77369-157-2

Golden Meteorite Press
103 11919 82 St NW
Edmonton, AB T5B 2W3
www.goldenmeteoritepress.com

We acknowledge the support of Canada Service Corps, TakingITGlobal, and the Government of Canada in promotional materials associated with the Project.

Thank you.

Table of Contents

Introduction

In late December of 2019, the World Health Organization (WHO) was notified of an outbreak caused by a pneumonia of unknown cause in China's Hubei province. The Chinese authorities insisted that human-to-human transmission was not evident, but the increasing cases in the country indicated otherwise. Cases began to appear in neighbouring countries, and by the end of January 2020, the WHO deemed coronavirus disease 2019, abbreviated as COVID-19, a global health emergency. The virus continued to spread across the world to over a hundred countries, growing in number and severity, making the WHO reconsider their label. On March 10th, 2020, WHO declared that COVID-19 had become a pandemic.

While many were unhappy with the lack of swiftness in handling the global crisis, past outbreaks put the lengthy decision period into perspective. Prior to COVID-19, the WHO declared the swine flu (or H1N1) outbreak in 2009 as a pandemic, and was met with criticism about the decision. However, with almost 35 million cases documented worldwide as of mid-August 2020, it is undeniable that the severity of COVID-19 far exceeds that of other outbreaks. The pandemic has led to devastating consequences: the loss of loved ones in isolation, worsening mental health, and economic hardships. For many, it is not an overstatement to say that life will never be the same as it was during pre-pandemic times.

What used to be simple outings to the grocery store or summer gatherings has become calculated activities associated with the risk of contracting a potentially life-threatening virus. As people have become more wary of each other and their surroundings, there is continuous demand for masks, sanitizing products, and other protective measures. During such unprecedented times, people have relied on news outlets and the Internet to stay updated and safe. While technology advancements have allowed people to communicate in ways once unimaginable, it has also resulted in misconceptions about science and the use of unreliable sources for information.

So what does salt have to do with misconceptions and COVID-19? This is the question that many readers will be wondering as they peruse this book. Yes, it is the salt that you are imagining—the salt used to season food, sprinkle on sidewalks, and relieve a stuffy nose caused by the cold. While salt is widely recognized for these uses, it has played a critical part in developing protective equipment for healthcare providers and the public. This book will delve into how and why salt became a multi-functional mineral that is used in both our homes and hospitals.

Salt has an extensive history that stems from its discovery by man, which is thought to be far earlier than any historical record. Without any processing, its natural form is called halite, otherwise known as rock salt. Many organisms need to ingest the mineral to perform metabolic processes and regulate fluid levels in the body. As you will soon learn though, humans have found uses for it that go far beyond our ancestors' imagination. It is commonly used in the kitchen as a seasoning agent and a necessary ingredient for food preservation of dairy and meat products. In the industrial world, it is bought by consumers to prevent ice formation in the winter or as a component of cleaning/laundry products. Given its widely accessible and inexpensive nature, it has a place in every household. Chapters 2 and 4 will take a look at the history of salt, and the mechanisms that explain why salt is useful in these contexts.

Most people share the knowledge that gargling with salt water can relieve a sore throat and runny nose—symptoms associated with the common cold. While scientific studies have shown that salt water can reduce the duration of a cold, they do not indicate that it is an effective method to prevent someone from contracting COVID-19. COVID-19 is caused by severe acute respiratory syndrome coronavirus 2 (SARS-CoV-2), which scientists call a novel coronavirus: it has never been seen in humans prior to the outbreak in 2019. The Centers for Disease and Prevention (CDC) emphasizes this by explaining that "the virus causing coronavirus disease 2019 (COVID-19), is not the same as the coronaviruses that commonly circulate among humans and cause mild illness, like the common cold".5 Given its unique properties, scientists have spent months trying to understand how the virus works, how it is spread, how it can be treated and more importantly, prevented. Chapter 4 explains how research shaped the scientific connection between salt and viruses and Chapter 8 discusses the importance of salt in the context of COVID-19.

Scientists have worked for months to understand the makings of the contagious virus, yet many people have searched for homemade solutions that claim to minimize the likelihood of contracting the virus. These do-it-yourself (DIY) solutions have been posted across the web for others to imitate, spreading the false notion that they are effective. One of these solutions, and the most pertinent to this book, is saltwater. Since salt is an easily accessible mineral, misinformation surrounding its use can quickly cause unintended and dangerous outcomes. In South Korea, a church in Gyeonggi province used the same spray bottle to spray saltwater into attendees' mouths without disinfecting it after each use. The church claimed that this would prevent the virus from infecting others, yet 46 of those who attended the service ended up contracting the virus in March 2020. This example alone demonstrates the danger of believing information that is not scientifically supported and assuming that knowledge can be applied to a seemingly similar situation.

As a result of social distancing measures, the public has relied on technology to communicate and educate themselves on current events both related and unrelated to COVID-19. One of the underlying themes in this book is the recognition of sources that provide misinformation. While it is important for all of us to stay updated on the current situation, news outlets, external websites, and even public figures are not always reliable sources. In late April of 2020, the president of the United States, Donald Trump, made a suggestion that disinfectant products could be injected to clean and clear the lungs from COVID-19. Out of concern for consumers' safety, Reckitt Benckiser—the parent company of Lysol—had to make a statement that their products were "under no circumstance" acceptable to inject into the body. While Trump posed the idea as a question, statements made by political figures can be taken out of context with dire consequences. Despite Trump's "sarcastic" intentions, the state of New York's Poison Control Centre observed a doubling in the number of calls for "bleach and Lysol-related issues" following his comment compared to the same timeframe a year ago.7 Although the spike in cases may not be directly related to Trump's comments, it does reflect the misbelief that many people have about the safety of using cleaning products to sanitize themselves.

Misinformation can be spread just as easily as a contagious virus. It is important to recognize scientifically supported information to make decisions that directly affect your health. As we continue to learn more about COVID-19, it's essential that we obtain information from scientific sources such as the WHO, and avoid the use of non-tested solutions when it comes to home remedies or other DIY prevention methods. Chapter 3 debunks some of the myths surrounding the use of salt in different applications and chapter 11 provides an overview of the DIY solutions that have circulated during the pandemic.

Finally, this book will acknowledge how salt is utilized within and outside of healthcare. During the pandemic, the primary focus has been on how to prevent the spread of the virus, boost our immune system, and protect ourselves. These are, of course, of utmost importance. But it's important to encourage the awareness of the research that has developed current prototypes of protective equipment, and acknowledge the time and effort devoted to the process. The public has to allow researchers to do their job and recommend ways to prevent the virus's spread rather than becoming self-made professionals. This book is an opportunity for readers to become educated on the scientific mechanisms that explain how protective equipment is effective. Salt, in particular, has a diverse range of medical applications that most people aren't aware of.

The most well-known application of salt in medicine is saline solution, a combination of salt and water with a concentration similar to our bodily fluids. It is administered intravenously (also referred to as saline infusion) often to replace lost fluids and electrolytes such as sodium. Most people do not know, however, that saline is also used in personal protective equipment (PPE) for healthcare providers. Chapter 14 discusses this topic in further detail.

Among the PPE, the public has shown a mass demand for masks. All types of masks have been used during the course of the pandemic: surgical, cloth, the N95 respirator. During the pandemic, it has been emphasized that N95 respirators should not be used by the public: they are incredibly valuable and difficult to acquire for healthcare providers. However, many have sought out N95 respirators with the belief that they are much more effective than any other mask. N95 respirators are meant to seal the area around the nose and mouth to maximize their filtration efficiency, compared to other masks used in clinical settings. That being said, meta-analyses—research that compiles and compares the results of several primary studies—indicate that N95 respirators do not have a significant difference in lowering the risk for contracting respiratory or viral illness compared to medical masks.

Since the severe acute respiratory syndrome (SARS) outbreak in 2003, research has been targeted at developing masks that both protect and prevent virus spread. More about salt's applications in relation to SARS as well as the H1N1 outbreak can be found in chapters 6 and 7, respectively. For instance, Professor Choi Hyo-Jick and his colleagues at the University of Alberta have been developing a salt filter since 2015 that enhances the protection of surgical masks. The formulation is meant to be used on the exterior of the mask and kill viral particles as they try to penetrate the mask. In their trials, the salt coating has been successful at deactivating, and killing several strains of H1N1. Chapter 10 describes the science behind salt filters and how they are effective, and chapter 9 explains how saline has been used in masks.

Researchers are also using salt's antibacterial properties to prevent the spread of viruses on frequently touched surfaces. Outbreaker Solutions, a research company, has been working with University of Alberta researchers to test the use of compressed sodium chloride (CSC) on surfaces such as door levers, hospital bedrails, and handrails. They are conducting research that will test whether salt can kill intestinal and respiratory viruses such as COVID-19. The project will hopefully add onto the findings of Choi and colleagues' who were the first to provide evidence that salt is capable of killing influenza viruses.11 Chapter 12 simplifies the processes of infusion and coating surfaces with salt, and chapter 13 expands on these concepts by specifically looking at door levers and handrails. Chapter 15 takes an in-depth look at how salt is being utilized to develop antimicrobial surfaces.

We've provided a glimpse into the array of topics that will be covered in this book. Already, it is evident that extensive research has been conducted with salt to enhance the protection of healthcare workers and the public. The intention of this book is to provide readers insight into why the usage of salt is supported in the aforementioned applications and comment on its relevance to combatting COVID-19. As you learn more about the myths surrounding salt, we encourage you to always verify information sources and be wary of misinformation. Most importantly, we hope you will come to appreciate what a pinch of salt is capable of accomplishing just as we did!

History of Salt

Salt, a simple chemical compound composed of sodium and chloride, is such a key player historically. From religious references to whole economies, salt has been an important part of humanity, life evolving around it. For thousands of years humans have looked towards salt for food preservation and salvation, countries looked towards salt for taxes, and salt became a symbol of purity for religions. Whole cities and towns were built based on availability of salt, and their proximity to sources of salt. In Britain, any town or city with the prefix "wich" indicates a place that is a source of salt, for example Norwich and Sandwich. Even going all the way back to the Egyptian Empire, they used a specific type of salt known as natron, and counted it as an extremely valuable resource. In fact, Natron Valley was a key region that supported the Empire, providing them with a source of salt. Through years and years of history, salt has been a constant factor of civilization. Once it was scarce and hard to isolate, but today it is available everywhere. From sea salt to iodized salt, it is one of the most basic ingredients in today's world, but this was only possible through the hardships of various civilizations.[13]

Before we dive deep into understanding the history of salt itself, we need to understand why it's so important. Salt, to a chemist is any cation ion and anion coming together to form an ionic compound, but to most others it is the white grainy substance used everywhere from our food to the roads. This white substance is actually just sodium and chloride ions held together, but what about it makes it so important to human health? Well, sodium ions (positively charged cations) are an extremely important part of extracellular fluid, helping to maintain the volume of blood plasma and ensuring normal cellular metabolism. But arguably even more important is the role sodium and chloride ions place in the nervous system. It is the change in concentration of these ions between the inside and outside of neurons that actually leads to formation of an action potential, the signal that neurons use to communicate. Even more, the chloride ions found in salt play an important role in the digestive system, as they're used to form hydrochloric acid (HCl). This acid is used by the stomach to digest food, and a key step in the digestive system. In fact, salt is not just extremely important for humans but for mammals and animals in general. Sodium deficiency in mammals can actually lead to cardiovascular collapse, acute blood loss, brain swelling and in severe cases, coma.[14] So, a proper amount of salt in one's diet is important and necessary for all humans, but how did something that was such a commodity in history become so easily available today.

Humans first started using salt when agriculture was developed. Normally, a person's salt intake was mainly maintained through the hunting and gathering diet, where the necessary salt was taken in by eating meat and fish. However when Neolithic humans started progressing to a more vegetable and grains based diet, a lack of salt in the diet was found. Along with this, it was also discovered that salt helps in the preservation of certain foods and meats. This eased civilizations dependency on seasonal based farming foods, and allowed for a year round supply of certain foods. The earliest record of the story of salt traces back to the Xia Dynasty in 2000 B.C, where the earliest known salting of fish occurred. Records also show the first known harvesting of salt occurring around 6000 B.C in the Chinese province of Shanxi. By 1800 BC, China had a process set up for acquiring salt from sea water, where sea water was boiled in clay pots until only salt crystals were left. In 450 BC, the original distillation methods were formed, where brine (salty water) was boiled in iron pots, leaving the salt behind and fresh water vapor then condensed to form drinking water. Around 225 BC, there are also records of the first drilling of brine (salt) wells. Salt also had a major influence on the food of that era, in fact around 500 B.C what we now know as soy sauce formed by preserving soybeans in salt, which later turned into a liquid and thus soy sauce.15 In 100 BC, salt had become such a commodity that Chinese emperors controlled its trade and used it to control the population. By 100 AD, half of China's revenue was from salt and its monopoly of the salt trade in Asia.15

The ancient Egyptians also used salt around 6050 B.C, with salt preserved fish and meats found in the tombs of rich pharaohs. During these times salt was thought of as a commodity, and thus only the rich could use salt in the afterlife. In order to actually get the salt, the Egyptians would use a process known as dragging and gathering. Salt, usually left behind from evaporated sea water, was gathered and bargained for, by everyone that could. This is one of the first records of salt being gathered, and shows its prominence in multiple civilizations. Similar to the Xia Dynasty, the Egyptians also used salt to create flavorful condiments, in particular oxalme was a popular one.15 Oxalme was made using water, vinegar and salt and is very similar to fish condiments of today. However, perhaps even more important than food the Egyptians had an alternative use of salt. Natron, what was known as divine salt, was used to preserve the pharaohs and those placed in tombs. It is a process very similar to the curing of meats and fish, but instead with the divine salt.

Around 2800 BC, the ancient Egyptians also traded salt, in the form of salted fish, to various Middle Eastern societies, the most prominent being the Phoenicians.15 They in turn started a salted fish trade in other parts of the Middle East, and by 800 BC, the Phoenicians and a few other Middle Eastern communities had started producing their own salt through North African lakebeds. Over time, the salt trade strengthened, from the very early civilizations such as the Xia Dynasty and the Egyptians to later the Europeans and North Americans, salt was a highly traded item, bringing cultures and people together.15

Salt was also of importance to the Romans. Roman emperors, by taxing salt, used it as a way to provide for their military campaigns and weapons during times of war. During times of political campaigning, or when the public's support was needed, the emperors would lower the taxes on salt. This became a common practice in many other countries later on.15 Also, recognizing the importance of salt, Rome itself was built near a place where salt is produced (a saltworks). Later in history, when saltworks were moved around, the Romans built a big road known as Via Salaria, also known as the salt road. This road became an integral part of the salt trade in Rome, and thus is known as Rome's first great road. Salt was so important to the Romans, that the word salary is actually derived from the word salarium, referring to salt.15 In fact, Roman soldiers were often paid partly in salt, hence the word salary, and it was thought of as a great honor to be given salt as payment.

In ancient times, salt was also used as currency, many times in fact to buy expensive objects or slaves. The phrase "someone not worth his salt" actually refers back to when slaves were traded for salt, and a "poor" slave wouldn't be worth his/her salt. Astonishingly, during the Roman and Greek Empires, an ounce of salt was often worth an ounce of gold, was traded accordingly. Throughout all of the Roman Empire, salt continued to be a commodity and something sought for, many countries fought and lost wars over salt.15
Various European countries were also a part of this world wide search, acquisition and trade for salt. Salt was being mined from salt mines in Austria by 400 BC in Salzburg, literally meaning salt town.15 Alpine salt miners, of Celtic descent, were also avid salt traders. Mining salt from various locations over Europe, they then traded this salt to the Roman Empire, Britain, Spain, France, North Africa and Turkey.15 A country doing particularly well during these times was Venice. A country formed of refugees, their main economic wealth came from salt works. Trading exclusively with Constantinople, Venice held a large monopoly over the salt industry during this time. It was when Constantinople fell to the Ottoman Turks, when the Venetian's eventually lost their monopoly and economic wealth. During medieval times, France

also rose to become a great producer of salt. One of the key features of this was the heavy salt tax levied in France, which was used to fund the French Revolution. Salt was an extremely important mineral throughout Europe over the ages, however its importance extended across seas.

Salt played an important role in the European exploration of North America. The first Native Americans, British explorers came into contact where in the Caribbean's, who at the time were harvesting sea salt. Later on, when explorers docked at Newfoundland, the French, English, Spanish and Portuguese used the sea water there, like the Aboriginals, to keep fish well preserved.13 This actually led to the establishment of a territory there, and British and French shipper men became the first official European settlers in what is now known as Canada. Later on in North American history, salt continued to be an important commodity. One of the items being heavily taxed by King George that led to the American Revolution, was salt itself. In fact, even during the revolution itself, the two armies tried to protect their own sources of salt, and cut off the opposing side's access to salt. It led to low salt access for the Confederation, and low morale for the rebels.13

Through the hundreds of years of human civilization, salt, just a simple ionic compound, has to lead to the triumph and fall of cities and empires. It has led to battles and wars, and a lack of it has resulted in devastating casualties. It is said that when Napoleon's forces retreated from Moscow, hundreds of troops lost their lives due to the side effects of a low sodium intake, mainly easy infection and low resistance to disease.17 But, although salt was once a highly sought after commodity, in the 21st century and for the last 100 years or so, salt has become available everywhere. Table salt is one of the most common household spices, and you would be hard pressed to find a household that doesn't have at least some form of salt, whether that be table salt or sea salt or another "organic" salt altogether. In fact, in recent years the opposite of history has come to unfold, where humans are said to be consuming too much salt. Research suggests that an over consumption of salt in packaged foods, and fast food has actually lead to certain health disadvantages namely heart attack or lower cardiovascular strength, and high blood pressure.14 So, even after all of these years spent on salt in human history, it continues to be at the center of debate and conservation, albeit in a different way entirely.

Debunking Salt Myths

Imagine this, you are cooking a meal for dinner; you cut the chicken and place it in the pan before gathering the rest of your ingredients: oregano, garlic, onion powder and chili powder. Your roommate reminds you that you forgot to take out the salt, which is when you say to them "I'm actually trying to cut down on my salt intake so I won't add any". This statement is one of the many myths surrounding salt that have been perpetuated by the media for the last decade. While the American Heart Association recommends staying in the healthy range of salt consumption, which is approximately 2300 mg a day, most of the salt you eat doesn't come from the salt shaker. What this means is that omitting to add salt to your home-cooked meals is unlikely to aid in maintaining healthy salt intake levels. This poses the question: "Well then where does my salt come from and how do I eliminate it from my diet?". The majority of salt intake actually comes from processed foods such as bread or cheese and unlike what most of us have come to believe, attempting to completely eliminate salt from your diet can cause more harm than good.

Salt, an essential nutrient that the human body cannot produce on its own, is needed to maintain a healthy blood pressure. The maintenance of blood pressure is a tremendously delicate balance that is primarily controlled by your kidneys, as they are able to alter blood volume. Consequently, blood pressure is altered through the use of hormonal mechanisms. Although many hormonal and neural mechanisms affect blood pressure, the most important hormones to understand concerning salt intake are renin and angiotensin II. Renin is released in response to a lack of sodium chloride, or salt, in the kidney. It then undergoes a series of steps that lead to the conversion of renin into the hormone angiotensin II.20 This process works to increase the reabsorption of sodium into the kidney. As more sodium enters the kidney, an increased amount of water enters the kidney via a process called osmosis. In this process, water will travel from an area of low solute concentration to an area of higher solute concentration. As a result, less water is excreted out of the body, thus triggering an increase of both the blood volume and blood pressure in the body. For this reason, a diet with no salt can actually cause higher blood pressure. The presence of excess salt in the kidneys that stems from a high-salt diet will trigger osmosis in the same manner, thus contributing to a higher blood pressure. If left untreated, the high blood pressure will continue to put strain on the vessels that lead to the kidneys which can cause them to malfunction. Given that the primary role of the kidneys is to filter toxins and unwanted materials out of the blood by excreting it into urine, its malfunction can result in serious complications and a build up of harmful toxins in the body.22 Although there are conflicting theories on the effects reducing salt has on blood pressure, the general scientific consensus is that the optimal levels of salt intake is around 2300 mg a day.

If you need any more compelling evidence that points towards how harmful attempting to completely eliminate salt from your diet can be, you need to look no further than the physiology of your muscles and nerves. Muscle contraction begins when a nerve sends a signal to the muscle fiber via a neurotransmitter. In order to trigger an action potential, which is what will cause the muscle to physically contract, the inside of the muscle cells need to be positively charged. Since sodium is an electrolyte and a positively charged ion, when it enters the muscle cell, the action potential will spread to the rest of the membrane.24 The pathway that ensues is a multi-step process that occurs in a matter of milliseconds involving calcium and energy (in the form of an organic molecule known as ATP) that eventually leads to muscle contraction. Without any salt in our body, it would be impossible to jumpstart this process and there has been evidence suggesting that a lack of salt can cause muscle cramps during exercise. This gives way to another common myth about salt intake: if you regularly exercise you should increase your salt intake to compensate for the loss of salt due to the physical activity. In reality, during an average 30 minute workout you actually lose negligible amounts of salt. Heeding the fact that the average American consumes 1100mg more than the recommended daily salt intake, it is entirely reasonable to assume that it is unnecessary to increase your salt intake on the days you workout.26 However, professional athletes, who regularly engage in multiple hours of high intensity physical activity every day, and manual labourers who work long strenuous shifts, can generally safely consume more than the recommended sodium intake.

In the recent craze of "clean eating", one trend that has garnered plenty of attention is the use of sea salt instead of table salt. However, despite the claims that sea salt is a healthier and natural alternative to table salt, studies have shown that this is simply a myth, as both salts contain 40% sodium. In fact, in most cases it is better to stick to table salt as it has been iodized. Similar to salt, iodine is also an essential nutrient that humans can only obtain through consumption, and it contributes to the adequate functioning of our thyroid glands as well as intellectual development. While iodine can also be found in seaweed, fish, and dairy, the amounts consumed through these foods are often not enough to satisfy the recommended intake. This is clearly evident among Europeans, who represent 20% of worldwide iodine deficiency cases as iodized salt is not a household staple in the continent.30 It was not until 1924 that including iodine to table salt became the norm in North America and even now iodine deficiency is the leading preventable cause of intellectual disabilities in the world. A study conducted by economists used American military data to examine the impacts of the introduction of salt iodization in the 1920's and the results were quite shocking. Men born in regions with low iodine levels before 1924 had an average IQ that was 15 points lower than the men born in this same region after 1924. The latter population was also significantly more likely to receive entry in the Air Force. While there is no problem with occasionally substituting table salt with sea salt for added flavour, if your main motive behind doing so is in the name of health, it may be time to rethink your choice of salt.

Over the past few years, the multi-billion dollar naturopath industry has created numerous gimmicks that appeal towards peoples' desire to improve their health in a completely 'natural' way. One such gimmick is something commonly known as halotherapy, which promotes the unfounded claim that inhaling salty air can boost immunity, treat respiratory conditions, and even clear skin.32 Halotherapy is a treatment offered at spas that involves spending 30-60 minutes in a treatment room that has been pumped with microscopic salt particles. Proponents of this treatment suggest that the salt particles can absorb irritants and toxins, thus removing them from your body upon exhaling, despite a lack of scientific evidence to prove the theory. As such, it is better to save your money and stick to massages and facials the next time you visit a spa.

Given all this newfound information I'm sure many of you are wondering: well then how exactly do you track your salt intake in order to ensure that you are consuming the appropriate amount? The most important thing to keep in mind is that 80% of salt intake in North America does not come from the salt shaker but rather from the foods you eat. In order to help combat the excess salt consumption that is rampant in Canada, consider the top six food sources of sodium and their alternatives. First on the list are baked goods, such as bread, muffins, cookies, and crackers. The amount of sodium in a slice of bread can quickly add up throughout the day, so replacing bread with corn tortillas or romaine lettuce is a great substitute. As for other baked goods, it is much easier to accurately track your sodium intake if you bake them yourself. Next we have fan favourite meals such as pizza, potatoes, and lasagna, which have especially high sodium content if you purchase the frozen dish. The easiest way to get around this is to prepare the meal yourself, but with busy schedules, that may not be a feasible option. In such cases, it might be a good idea to limit the amount of frozen and processed food you eat. For a quick meal you can make a pizza on a pita bread rather than regular pizza dough, which will not only save you time, but also cut down your sodium intake. The third culprit is processed meats, which includes sausages, deli meat, and burgers. If there's no low sodium options available for these foods, your best bet is to find fresh meat and cook it within a few days or to choose meats that are baked, grilled or roasted. The next item on the list, cheese, is easy to replace with a low sodium substitute by choosing mozzarella cheese which contains less sodium and calories than most types of cheese. Following cheese is soup which surprisingly can easily put you over the daily recommended salt intake. Some low sodium options include: Health Valley Organic No Salt Added Chicken Noodle, Amy's Organic Light in Sodium Lentil Vegetable, Health Valley Organic No Salt Added Minestrone, and Pacific Natural Foods Organic Light in Sodium Creamy Tomato. The last product on the list are often overlooked sauces; soy, teriyaki, and barbecue sauce contain a lot more sodium than you would expect. There are plenty of low sodium sauce recipes out there but if you're looking for something you can pick up at the store be sure to check if your local grocery store has low sodium options for your favourite sauces. A general rule you can be sure to follow in order to lower your overall sodium intake, is to choose products that have a sodium content lower than 15% of the daily value.

A thorough evaluation of the effects of salt intake on physiological health reveals the importance of moderation in daily consumption, as well as an increased need for accurate nutritional information. While a complete elimination of salt from our diets can result in a plethora of complications, an excess can be similarly detrimental. In an age where one has access to an overwhelming amount of resources, we must make an effort to seek out academically sound advice as advocates of modern science and healthy living.

How does salt work?

An entire chapter dedicated to how salt works? What could possibly be written here that I don't already know?

This is generally the first thought (and the first misconception) that pops into most minds while reading the chapter title however, what we cover here is just the tip of the iceberg; for there is an immense amount of scientific jargon related to salt that could be discussed in future books, and more scientific discoveries yet to be made on the benefits and detriments of salt. This chapter may seem slightly arbitrary at first, however it will shed light on the various applications of salt from a scientific, health and lifestyle, and medical perspective.

From a chemical standpoint, a salt is an ionic compound that is the result of a neutralization reaction between equal amounts of an acid and a base. Acids are substances that release positively charged hydrogen cations (H+) when in water, while bases release negatively charged hydroxide anions (OH-) when in water. The reaction between a cation and anion is what produces the neutralization reaction with a neutral charge of zero. In this case, the H+ cation and OH- anion neutralize each other and produce a generic salt. The properties of salt differ based on what two ions are combined. Common table salt, referred to as sodium chloride (NaCl) in the world of science, is an ionic compound that involves the ionic bonding between a positive sodium cation (Na+) and a negative chloride anion (Cl-). Sodium is a reactive metal and given the opportunity, will react with sweat to form sodium hydroxide, a known corrosive substance. The element chlorine is a yellow, corrosive gas with toxic characteristics that can cause acute damage to the upper and lower respiratory tract, and should not be directly inhaled. When these two substances react together through the chemical reaction of 2 Na (s) + Cl2 (g) 2 NaCl2 (s), it produces the most common form of salt, sodium chloride. Properties of salt range across a spectrum and depend on what two ions are combined in the reaction. Pure sodium chloride is a white crystalline solid however, it may have a slight blue or purple tinge to it due to impurities during packaging. Water and its polarity attracts the positively charged sodium cation and the negatively charged chloride anion, allowing table salt to dissolve in water, as the covalent bonds of water are stronger than the ionic bonds in salt. This interaction is commonly referred to as an ion-dipole interaction. The resulting mixture is a homogenous solution, where the solute (water) and the solvent (salt) are equally distributed across the solution. The chemistry behind salts and its uses in the world of science are never-ending, and have provided scientists with an immense amount of information that continues to aid in further researching the various roles salt plays in the chemical sciences.

Shifting the focus to a nutritional perspective, salt and its usage in nutrition dates back thousands of years, and is a focal point when considering health and lifestyle choices. From a flavour enhancer to its other lesser-known roles in food preservatives, the various functions of salt as an essential nutrient proves its versatility and multifaceted roles in human nutrition.

According to an article authored by the Institute for Quality and Efficiency in Health Care (IQWiG), the 5 basic qualities of taste include sweet, sour, salty, bitter, and savoury (also referred to as umami). In low concentrations, salt aids to reduce bitter tastes and enhance sweetness in food as the sodium ions suppress the bitter flavours to make the sweeter flavours seem stronger. In high concentrations, salt suppresses sweetness and enhances umami and savoury flavours.

Salt is an effective food preservative as it reduces water activity by drawing water out of foods via the process of osmosis. Osmosis is the pathway in which water moves across a cell membrane, and equalizes the salinity on both sides of the membrane. If enough salt is added to a product, the water activity is lowered enough to inhibit bacterial growth and disrupt the proliferation of microbes that are known to cause diseases and food poisoning. Several disease-causing microbes are also unable to grow in the presence of salt. It has also been suggested that "for some microorganisms, salt may aid in limiting oxygen solubility [and can] interfere with cellular enzymes" further stunting their growth in food products . With modern advances in technology, such as the creation of refrigerators and freezers, few foods are preserved solely by the addition of salt however, it remains as a commonly used component for creating an inhospitable environment for pathogenic organisms to survive in.

As a texture aid, salt has the ability to strengthen gluten structure in yeast-based breads, enabling the dough to hold carbon dioxide and subsequently allows it to rise when baking. For cured meats, it improves its tenderness and retains moisture by breaking down proteins and muscle fibers within the meat. Salt, in a lesser-known role, can also be thought of as a colour enhancer for processed meats as it prevents these products from turning gray.

In one of its most important roles as a nutrient source, salt helps with fluid balance, transmits nerve impulses, contracts and relaxes muscles, and maintains the proper balance of water and minerals in the body. According to an article published by The Harvard School for Public Health, it is "estimated that we need about 500 mg of sodium daily for these vital functions". Salt is the dominant source of both sodium and chloride in the human diet. As a vital source of electrolytes, chloride ions regulate blood pH and blood pressure. These ions are excreted through sweating or via urine, and must be replenished through the diet. However, this poses the question, how much salt is too much salt?

The average amount of salt an individual should consume is largely based on weight and age. For the average adult, the amount averages to 2300 mg of sodium per day. According to a study published by Health Canada in 2017 titled The Sodium Intake of Canadians, the average Canadian adult consumes an estimated 2760 mg of sodium per day, with males consuming more sodium than females at an alarming rate of 90% excess sodium between the ages of 14-30 years old. Health Canada recommends Canadians to limit their sodium consumption to 1500 mg per day. Diets high in sodium can consequently lead to high blood pressure, which is a major risk factor and predictor for developing heart diseases and strokes. Heart disease and stroke are the leading cause of death in Canada, with 1 in 4 Canadian adults diagnosed with hypertension (high blood pressure).

In layman terms, the amount of salt consumed correlates with the amount of water the body retains; more salt consumed, more water retained. This excess water raises blood pressure and can adversely affect other vital organs including a larger strain on the heart, kidneys, arteries, and brain.

The heart is the primary organ in the circulatory system that is responsible for pumping blood throughout the body, supplying oxygen and nutrients to other organs and tissues, and removing carbon dioxide and other wastes found in the system. An increase in sodium levels in the body draws water into blood vessels, increasing the overall blood volume. A surge in blood flow due to more strenuous pumping by the heart puts more force on the blood vessels, raises blood pressure and overtime, may stretch blood vessel walls, and cause a buildup of plaque. Plaque buildup within the arteries is commonly referred to as atherosclerosis and is composed of fat, cholesterol, calcium and other substances found in blood. Plaque hardens and narrows the arteries as blood flow is decreased and can increase the risk of strokes, heart attacks, and heart failure.

The body excretes waste products and unwanted fluids via the urinary system that uses kidney filtration as the primary medium. These waste products and fluids are drawn out of the kidneys through the process of osmosis and are excreted as urine. A high salt diet alters the balance between sodium and water, causing the kidneys to have a decreased function in regulating the amount of water being drawn out and consequently, results in high blood pressure. This places an immense amount of strain on the kidneys as they are unable to filter blood, leading to kidney diseases. If kidney disease is left untreated and the high blood pressure is not correctly managed with medications and lifestyle choices, the damage can lead to kidney failure where the body will slowly become poisoned by its own toxic waste and unfiltered products. Additionally, a higher salt intake can increase the amount of protein in the urine, which is a major indicator and risk factor for kidney diseases.

According to the World Health Organization (WHO), one of the most cost-effective measures countries can recommend is to reduce overall salt intake to drastically improve population health outcomes.

While there have been a number of studies and research conducted on the correlation between salt, high blood pressure, and heart or kidney diseases, studies have also linked levels of salt intake affecting overall brain health however, the "mechanisms involved are poorly understood". A study led by Dr. Costantino Iadecola and his team at Weill Cornell Medicine explored the correlation between high salt diets and brain function. Mice were fed a diet that contained up to 16 times more than the recommended amount of sodium, as this is comparable to human excessive salt intake. After eight weeks, the results were startling; the brains of these mice showed up to a 30% reduction in blood flow in comparison to the control group that was fed a normal diet. Researchers also noted that the mice with a high salt diet had "difficulty in three cognitive tasks: recognizing objects, navigating a maze, and building a nest". It was hypothesized that blood vessels within the brains of mice fed a high-salt diet did not dilate as needed when prompted to, reducing overall brain blood flow. Nonetheless, when these same mice were returned to a normal diet, both blood flow and overall cognition improved, which suggested that negative effects of excessive salt consumption could be reversed overtime. This has been thought to be applicable to humans as well, where sodium intake drastically affects brain activity as tissues are not receiving enough oxygen and nutrients to function optimally.

On the other hand, a lack of sodium in the diet can potentially lead to a dangerous condition known as hyponatremia. Sodium levels deteriorate through perspiration (sweating) and urinating however, it is not enough to cause a sodium deficiency that is detrimental to the human body. With that being said, it is possible to become sodium-deficient through overexertion and severe malnutrition. On the flip side, sodium levels can be depleted by drinking too much water and diluting its overall concentration in the body. Hyponatremia is the extreme loss of sodium within the body that can trigger negative symptoms that range from muscle cramps, nausea and vomiting to shock, coma, and death. This is the reason why athletes are routinely advised to consume electrolyte-rich sports drinks that aid in increasing sodium levels and other essential minerals that are depleted during strenuous activities.

From ancient history texts and scriptures, salt has been thought to possess healing properties. From salt water gargling for soothing a sore throat to saline solutions to heal wounds, salt does more than just making foods more palatable. Are these just ancient myths or can these claims be supported by scientific evidence and research?

"Rubbing salt in a wound" is a very literal, stinging phrase. However, just because it was a viable antiseptic in past centuries, does not mean it still is today. Salt will sting an open wound and cause a considerable amount of pain due to its rough texture, and can further cause a wound to tear. Therefore, wound cleansing with saline helps to heal the environment and decrease the potential for infections. Saline is a mixture of salt in water and in different contexts, has different concentrations of salt in the solution. Normal saline solution has a salt concentration similar to what the human body produces as tears and other bodily fluids (0.9% saline). This solution will not sting or burn when applied to a wound and can be used in a number of different ways including (but not limited to) rinsing contact lenses, nasal irrigation, and bladder irrigation. According to the Journal of Athletic Training, normal saline solutions are gentle on skin, do not irritate the skin, do not further damage healing tissues, and either adds or takes fluid from the wound bed. It does not alter the bacterial flora responsible for repairing damaged skin. As an overall, normal sterile saline solutions are the most appropriate and preferred cleansing solution used by medical professionals as it is a nontoxic solution that does not damage healing tissues, and is effective with minimal pain. So, does that mean salt water and saline are the exact same?

The short answer - no. Solutions of saline found in hospitals and the salt water found in the ocean may seem similar however, there are key differences between them that makes saline the optimal choice when dealing with wound irrigation. Medical saline has a lower concentration of salt in its solution compared to seawater. Additionally, medical saline is sterilized in order to eliminate and kill dangerous bacteria, viruses, and other potential organisms. Whereas sea waters are home to various microorganisms including bacteria, algae, fungi, and more. When these microorganisms enter an open wound, they have the potential to cause severe inflammation and infections, slowing the healing process and further damaging tissues.

Other than wound irrigation and cleansing, salt can help regulate other bodily conditions and diseases. According to the National Eczema Foundation, bathing with a quarter cup of salt can ease symptoms of flare ups and itching. Salt water mouth rinses work to reduce dental bacteria by creating an alkaline environment with an increase in pH levels in which bacteria and other organisms cannot survive. The use of salt water from a dental perspective also promotes healing and can be used after minor dental surgeries to help the mouth recover.

With advancements in technology comes advancements in healthcare options, and many are now opting for non-medicative routes to alleviate and reduce symptoms for various conditions or diseases. A new form of alternative medicine called Halotherapy is continuing to rise and is marketed as a salt-based treatment that is a "non-invasive, drug-free, chemical-free, all natural solution". Halotherapy is derived from the Greed language; "halos" in Greek translates to "salt". According to the Salt Suite , a company that provides Halotherapy services, Felix Bochkovsky was a Polish health official in the mid-1800's that observed how salt mine workers rarely suffered from any respiratory ailments and diseases. He hypothesized that inhaling the salt aerosol on a daily basis prevents these respiratory diseases and other illnesses. With his discovery, those with asthma, allergies, emphysema, and other respiratory ailments went to the salt mines and speleotherapy (salt cave therapy) was created. The environment in salt mines have no pollutants and a stable temperature, and these elements work together to create an optimal healing environment. While halotherapy is a fairly new concept in North America, it has frequently been used in countries across Eastern Europe. The unrefined rock salt that is used in halotherapy included other mineral salts in varying concentrations that have been credited with having some type of therapeutic properties. With the use of a Halogenerator, companies have been able to reproduce a similar therapeutic environment comparable to salt mines and help reduce symptoms for those who suffer from respiratory illnesses, skin conditions, and more. Halotherapy is advertised as the simplest, non-invasive medication where clients are able to simply sit back and relax. The benefits of halotherapy are listed as:

- Reduces symptoms of common allergies, asthma, and respiratory issues
- Improves physical endurance and enhances performance in sports
- Alleviates symptoms of skin rashes, eczema, psoriasis, and other skin conditions
- Clears up nasal cavities and sinuses to enhance breathing

- Relieves overall stresses and tensions

The benefits and effectiveness of halotherapy continued to be researched and examined from more scientific perspectives to support these claims.

As an overall, salt does more than just enhance the flavour of foods. While some of its uses are unknown or unfamiliar to most of society, salt and its various applications continue to amaze us in ways we didn't think possible.

Salt and its Uses

Salt is that white substance on the kitchen counter, you can find in every household. From iodized to kosher, there are so many different varieties of salt, but in the end they're all just NaCl, with a couple of tweaks. Salt has been in use by humans for thousands of years, it has caused wars and countries to split apart. It has both built and broken economies. But nowadays it is in everything that we eat, from chips and salsa to frozen pizza. You'll be hard pressed to find something at the grocery store that doesn't contain salt (except maybe fresh vegetables and fruits). We use salt to enhance the flavors of the food we eat, and this mainly stems from the fact that as humans 'salty' is a key taste that almost all people can experience and enjoy. But there's various other purposes of salt as well. In the winter, in countries such as Canada, America, and Russia salt is used to help the snow on roads melt faster1 . Salt has also been linked to have certain effects against viruses and bacteria, just like humans and animals too much salt can be deadly to them as well . But in addition to all these purposes, salt is an extremely important part of normal bodily function for human beings. In fact, in the past humans had to look for and actively sought out salt, as low sodium intakes would lead to deadly ailments. Nowadays, the opposite is being seen, where too much sodium intake also shows a health detriment . However, whether one is taking too much or too little salt, or is using it for its many other purposes, it is important to understand how it works, so that you can responsibly use, and eat, it in the future.

One of the oldest purposes of salt was food preservation, and till today salt is still repurposed in preservatives for this. There are two main mechanisms in which salt is used to preserve food, and that can be through curing or brine. The first method involved granules of salt whereas the latter involves a water-salt solution (known as brine). In curing, specifically dry curing, granules of salt particles are spread across the outside of meats and certain vegetables, allowing them to stay fresh and not spoil for weeks at a time . Some of the most historically important cured foods include cured fish and cured meats such as turkey and bacon. There have been references of cured fish dating back to at least the early Egyptian times . Vegetables such as beans and cabbage were also dry cured back in the day. Using brine is another way to preserve foods, and typically is when meats or vegetables are placed in an approximately 20% salt-water solution. These methods were extremely useful back in the day as they allowed for people to travel longer distances without being limited by fresh or close by food sources. It also helped tribes and communities survive harsh winters and long famines, as a food store meant there was at least a little to eat at all times. But, by placing them in a brine solution, for example which has much more solute in the solution than inside the vegetable, the water inside will be drawn outside. The imbalance of solute will cause the water to go from inside the vegetable into the brine solution, effectively dehydrating the vegetable6. Similarly, when meats are cured, the layer of salt on the outside will draw the moisture out from the meat, and dehydrate it. These processes were very important in places where the sun could not be used to dehydrate the food, in order to make it last longer . Moreover, dehydration of the food is extremely important, as most microorganisms and bacteria, cannot survive without water nor live long in extremely saline environments for long. Thus, by curing and salting foods, the spoilage of food would be postponed, for weeks sometimes even months, which is why these methods are still seen used in indigenous tribes all around the world today60!

One of the most interesting applications of brine and using salt as a preservation, is when salt is used in fermentation. Fermentation is the process used to pickle cucumbers and cabbages, while also increasing their shelf life thus effectively preserving them. In fermentation, the vegetables are placed in a low concentration brine, usually 3-7 percent brine . When the salt concentration is this low, a lot of bacteria are killed but not all. Lactobacillus are one type of bacteria that thrive in a 3-7 percent brine, and these are what effectively 'pickle' the vegetables. Once the vegetable is placed in the brine, the lactobacillus will start to grow and suppress other bacteria by using up the resources and causing an acidic environment. The main process which the bacteria use is lactic acid fermentation. In this, the bacteria starts breaking down the sugars found in the vegetables for energy, and produces lactic acid and mild alcohol as a byproduct. It's the lactic acid plus the salt from the brine that causes the vegetable to pickle, and gives them their characteristic tangy and acidic taste61. This process also helps preserve the vegetables for extended periods of time, sometimes they can last even months without refrigeration.

The whole method of pickling can take anywhere from 2 weeks to multiple months depending on what you are trying to pickle, what the concentration of the brine solution is, and how tangy/acidic you want your vegetable61.

In today's day and age another important purpose of salt, is its unique ability to help us get around safely in winter. What I am referring to here is salts ability to lower the freezing point of water. In winter, normally water freezes into ice at 0 degrees Celsius . At this temperature there is usually a thin layer of water on top which is in equilibrium with the ice below, such that some water is constantly turned into ice, while some ice turns into water. When salt, for example table salt NaCl, is spread onto sidewalks and roads in the winter, it comes into contact with the water and starts dissolving. This forms a water-salt solution, aka brine, and brine has a lower freezing point than normal water. So brine will freeze at -21 degrees Celsius whilst normal water freezes at 0 degrees Celsius. As more salt comes into contact with snow/ice, more of it will melt into water, and as more water is formed more salt particles will be dissolved, forming a repeating cycle. But the reason that salt can lower water's freezing point, has to do with the chemistry of the molecules62. Ice molecules are very rigid, and have a very stiff structure, and they need to be able to maintain this structure in order to stay compact and 'icy'. However, water molecules are more flexible, and thus do not require such a rigid structure. So when a salt is added it will dissociate, for example when table salt is placed in water it will split into two ions, sodium (Na) and chloride (Cl). These ions will then interfere with the bonds the water molecules can make to each other. Thus, in ice, the salt molecules will cause the bonds to break and weaken causing the ice to effectively melt and turn into water/slush (depending on temperature and brine concentrations)62. This effect is only useful till a certain extent, usually around -20 degrees Celsius as this is when brine itself will start to freeze. To increase the efficiency of this process, it has been found that different salts such as CaCl2 work better at lower waters freezing point. This has to do with the fact that CaCl2 will dissolve into 3 different ions once it comes into contact with water (1 calcium and 2 chlorine), and the more ions present the more they can interfere with and cause bond breakage in ice62. As humans, this process is extremely important to us as it makes the roads safe to drive in the winter, but the side effects of using salt can be extremely detrimental to the environment. Chloride is considered a toxin to aquatic animals and plants, and can lead to downfall of multiple food webs. This makes it extremely important to sustainably use such processes and to understand how they work, to make them much more environmentally friendly62.

Historically and in recent years salt has been very important, but salt has been playing an interesting part the last couple of months as well. As the pandemic continues on relentlessly, researchers are looking at other strategies to combat viruses. One of these methods involves using salt on masks, to help potentially deactivate viruses and provide an extra barrier of support. What most people don't realize while using masks is that viruses can actually survive for hour's maybe even days on surfaces, including gloves and masks52. And the thing is, people have a tendency to touch their face, including touching the mask and then touching a non-protected area. So even when we are trying to be safe, it might be the simplest things that cause a detriment to our health. This is where the idea to 'deactivate' viruses once they are on masks stems from. The masks have a salt coating on top of them, which act to kill the viruses. So when viruses carrying micro-droplets of water come into contact with the mask, the salt will dissolve into the water droplet2. The water will start to evaporate leaving only the salt particles behind, which begin to grow with extremely sharp edges. These edges work to puncture holes into the walls of the viruses, causing them to deactivate or die52. This new technology could be a game changer for the second wave of the pandemic, lowering the burden on society as a whole.

Salt, a simple white substance used in everything from food preparation to on the road, has so many purposes and utilities. From some of the earliest human civilizations to modern society, the importance of salt has only grown55. We used to use it for food preservation, and as a symbol of trade and exchange. Nowadays we use it for safe road conditions in winter, and to act as a defense mechanism against viruses52. As research continues to grow, so do the uses of salt, and that is why it's important to understand how salt works. It lowers the freezing point of water, and deactivates viruses along with killing bacteria and dehydrating meats/vegetables61. Hundreds of uses, and as modern society ages we continue to understand how and why salt works, along with all of its uses.

The Link between Sodium Levels and COVID-19 Mortality Rate

The COVID-19 pandemic has resulted in a surge of scientists researching the structure of the virus, potential vaccines, short-term and long-term effects of the virus, as well as the pre-infection lifestyle habits that can contribute to severity of symptoms. A recent study published in the European Journal of Internal Medicine found that there may be an unexpected link between low sodium levels and a higher mortality rate for those infected with COVID-19. Due to the global nature of the world's economy, the virus, which originated in China, quickly spread to many other countries. As a result, the sum of the number of deaths in these countries has long surpassed that of China. In fact, the mortality rate of the virus is nearly three times lower in China than the average rate in other countries.64 Although there are many factors that may have affected these statistics, such as lifestyle, prevalence of wearing masks, social distancing laws, and genetics, another factor to consider is diet. China is a leading country for salt consumption rates throughout the world, with the estimated average adult consumption being 10.9 g. This amount is more than double the five gram limit recommended by the World Health Organization to maintain healthy salt consumption levels. As previously mentioned, the majority of salt intake does not come from salt added to food; however, this finding only applies to the Western world. In China, the added salt in traditional Chinese meals far outweighs the salt consumed via processed foods.

The family of coronaviruses that can cause disease in humans are called human pathogenic coronaviruses. The virus binds to the cell they are targeting through something known as the angiotensin-converting enzyme 2 receptor, or ACE2 receptor. This receptor is found in specific epithelial cells, which serve as a barrier between the inner and outer parts of your body. The epithelial cells that carry the ACE2 receptor line the kidneys, blood vessels, lungs, and intestines. Studies on animals have shown that the expression of the ACE2 receptor is suppressed when one has high dietary sodium intake levels. This means that if an individual regularly consumes high levels of salt, the receptor that coronaviruses bind to will likely not be expressed as much compared to those with lower sodium levels.68 As a result, it is possible that they are less susceptible to contracting the virus.

A study conducted by Adrian Post and others hypothesized that decreased sodium levels could cause damage to cells and heighten the risk of someone developing a deadly infection if they catch COVID-19.68 However, the ACE2 receptor is also responsible for anti-inflammatory properties and can protect against acid aspiration-induced Acute Respiratory Distress Syndrome. This syndrome is a major lung condition that leads to a decrease in the amount of oxygen in the blood. Therefore, if one has very low sodium intake and is infected with COVID-19, the receptor to which the virus attaches would hypothetically provide protection against inflammation and acute respiratory distress syndrome, while still leading to a potentially more fatal infection. Other factors that have been shown to affect the expression of the ACE2 receptor include anything that can cause the sodium levels in the body to become unbalanced. These factors include diarrhea, vomiting, and perspiration, as these would all lead to a sudden loss of sodium in the body. This would result in the same effects as consuming low amounts of sodium as it changes your sodium balance in the same manner.71 For this reason, it may be beneficial if healthcare workers monitor severe COVID-19 patients' sodium intake levels and work to promote higher levels by providing high concentration saline solutions through an IV. As for the general population, if you are thinking of severely restricting your sodium levels anytime soon, you might want to wait until the COVID-19 outbreak is under control. However, if you are someone that has been told to keep your sodium levels low by a medical professional due to a condition such as diabetes, hypertension, or kidney disease, it is important to not make any changes to your diet without first consulting your doctor.71 As mentioned in a previous chapter, it is imperative to note the overwhelming amount of evidence that suggests very high or low salt intake levels can lead to numerous health complications.

There is one method to relieve COVID-19 symptoms that has been around for centuries and you do not need to consult your doctor before attempting it: gargling salt water. Scientists at the University of Edinburgh in Scotland are currently conducting a study under Professor Aziz Sheikh to determine if a salt water treatment could alleviate COVID-19 symptoms. The sample population tested are individuals who had been either confirmed or suspected of having the virus. The trial was based on the researcher's pilot study published in the Journal of Global Health that discussed if saline nasal irrigation and gargling should be advised as part of a COVID-19 treatment regimen.72 There were 66 individuals studied, all of whom had upper respiratory tract infections due to various infectious viruses, including coronaviruses. Those who were told to use saline nasal irrigation and gargling were to do so at their discretion up to a maximum of 12 times a day. The study found that this variable group reported their duration of symptoms was reduced by an average of 2.5 days in comparison to the control group that did not utilize the saline solutions and followed routine treatment. Furthermore, the control group took over the counter medications at a rate that was 36% higher than the variable group.72

The proposed reason for this improvement lies in the histology of epithelial cells, which we discussed earlier in the chapter. It is within these cells that antiviral effects come about by the formation of hypochlorous acid (HOCl), the active ingredient in bleach.72 Bleach has been proven to be an effective disinfectant against all types of viruses and hypochlorous acid is produced from chloride ions, which can be found in salt (NaCl). Thus, the study was testing the hypothesis that the supplied salt in the saline solutions inhibited the virus by providing the necessary chloride ions to produce hypochlorous acid within the cells. All of the viruses tested were inhibited in the presence of NaCl, thus providing statistically significant evidence that the hypothesis is correct. When only the individuals affected by coronaviruses were singled out and studied, it was found that their symptoms were reduced by 1.7 days on average compared to the control group. The data presented in the study supplies evidence that salt has an antiviral effect that can potentially be implemented as part of the routine treatment plan for coronaviruses.73

Despite the rather new information that salt water may be able to fight COVID-19, its use to alleviate symptoms of the common cold are nothing new. Some of you may be wondering how this household ingredient has the power to improve your sinus concerns and the answer is less complex than you would think: osmosis. Osmosis describes the phenomenon of water moving from a place of low solute concentration to a place of high solute concentration. In this case, the solute is salt and the solvent that it is submerged in is water. When you gargle salt-water, you submerge the inflamed tissue, which contains excess fluid, in the salt-water. When this occurs, the salt concentration in the tissue is lower than in the saline solution you are gargling with. As a result, the excess fluid within the inflamed tissues comes to the surface of the throat, bringing along any bacteria or viruses with it.74 Upon spitting out the water, you are able to get rid of those viral particles and bacteria; however, this does not cure the virus, it simply alleviates symptoms. Furthermore, any bacteria left within the tissue has now become dehydrated and if it does not maintain an adequate supply of water, it will eventually die. Another reason why your throat may feel better immediately after gargling with salt-water is because it acts as a lubricant for your throat, thus helping with irritation. Other benefits include the ability to restore the pH balance in the throat by neutralizing acids, washing away mucous, and increasing blood flow to the throat. The latter occurs when the capillaries, or tiny blood vessels, in the throat dilate, thus allowing for more blood to circulate and reach the throat. This also leads to faster circulation of the blood cells that are fighting off the virus.75 To reap the benefits of gargling salt-water, it is necessary to mix the correct ratio of the simple ingredients. It is recommended that 200 mL of warm water is used, as the heat will dissolve the salt easily, along with two teaspoons of table salt and one teaspoon of baking soda. The baking soda assists in ridding your throat of any built-up mucous. For the best results it is advised to gargle two to four times a day consistently while you have symptoms. When doing so it is necessary to drink a lot of water to ensure you are not drying out your other cells. Although this is generally a safe treatment for most, it may not be the best option if you have high blood pressure. In this case, drinking home-made chicken soup (store-bought alternatives have high amounts of sodium) can result in similar effects.

Nasal saline irrigation is another method that can be helpful in treating symptoms of the common cold that was tested alongside gargling salt-water in the aforementioned studies. The saline solution is typically delivered using an irrigation device such as a neti-pot, or occasionally a nasal spray bottle. However, the latter does not produce benefits to the same extent that an irrigation device would, since the fine mist simply moisturizes dry nasal passages. Sinus rinsing with an irrigation device can help break down thick mucous while simultaneously removing debris, dust, bacteria, and pollen from the nasal passages. This can aid individuals suffering from allergies, chronic sinus issues, the flu, and even COVID-19.[79] It is important to not perform this nasal rinse with water without salt added as it can lead to irritation, and it is best to use previously boiled or distilled water to make the saline solution. Generally, most nasal irrigation devices work the same way and, if used properly, are not overly uncomfortable. Although there might be slight variations between specific products the most common way to use one is by leaning over a sink, tilting your head sideways, and then inserting the spout into the upper nostril and allowing the saline solution to flow through and exit via the lower nostril. You then repeat these steps with the other nostril.[79] However, be sure to look at the specific instructions on your nasal irrigation device before using it and if there's no instructions you can always ask your pharmacist.

As contemporary research continues to suggest a correlation between low sodium levels in the body and a higher susceptibility to contracting viruses, it might be advantageous for us to implement salt in our treatment plans for COVID-19. While household remedies such as salt water treatments and nasal saline irrigations have been around for a long time, new studies and patient experiences have once again validated their efficacy in mitigating viral symptoms.

Use in H1N1 outbreak

In the last chapter, we took a look at how salt was used in the SARS outbreak. This chapter will be discussing the H1N1 outbreak, and how it shaped research for the next decade to come.

In the spring of 2009, the Centers for Disease Control and Prevention (CDC) announced that there were cases of influenza A H1N1 virus in California, US that originated in Mexico. The term influenza refers to respiratory disease—illness affecting our ability to use our lungs and breathe—that is caused by type A, B, or C viruses. Type A viruses are the most dangerous of the three because they can cause severe health complications in humans (unlike type C viruses that cause mild illness). They can also infect both humans and animals (unlike type B viruses that only affect humans), and have been linked to pandemics in the past (unlike type B and C). Outbreaks of swine influenza A viruses are common among pigs, hence their name "swine viruses", but it was the first time that humans had contracted a strain of the viruses. While there remain conflicting opinions about its origins, research seems to support the theory that the virus came from Mexican swine.

The H1N1 virus causes influenza symptoms that can include the following listed by the CDC: "fever, cough, sore throat, runny or stuffy nose, body aches, headache, chills, fatigue, nausea, diarrhea, and vomiting". While fever is listed as the first symptom in the list, a significant portion of infected individuals do not display fever as a symptom. The contagious period for influenza typically starts the day before an individual shows symptoms, and can last up to a week, however those who are more susceptible to illness (and have a weaker immune system) may have a longer contagious period. This period is part of virus shedding, which refers to infected individuals who release the virus into the environment (through coughing or sneezing). What's important to know about virus shedding is that it can last up to 6 weeks without necessarily infecting healthy individuals. Scientists are still investigating the length of the contagious period for COVID-19, however it's estimated to start two or three days before the onset of symptoms, up to at least ten days after they've appeared.

A key similarity between H1N1 and COVID-19, is that infected individuals can spread the virus even when they do not show any symptoms. Both are transmitted during person-to-person contact (through coughing, sneezing, or talking) that causes the virus to enter the mouth or nose of an uninfected individual. It's also possible for someone to become infected by touching an infected person/surface, and then touching parts of their face. A study by Wölfel and colleagues suggests that SARS-Cov-2 is not active in the stool, blood, or urine of someone who has the virus: it is infectious in throat and lung samples. Infected individuals often cough up sputum, a combination of mucus and saliva secreted by the respiratory tract. In Wölfel's study, sputum samples demonstrated proliferation, or viral growth, which indicates that it can infect healthy individuals through contact with their eyes, nose, or mouth. COVID-19 is considered more contagious than H1N1 because an infected person is able to infect several others with tiny respiratory droplets unlike the flu.

The longer asymptomatic period with the COVID-19 virus makes it more difficult to assess whether someone is infected. As a result, people who are unaware of their infection will easily spread it to other people, and touch surfaces that will eventually infect others. So it's not surprising that there have been clusters of COVID-19 cases (where the infected individuals do not live in the same household) connected together by a specific place. The sheer number of those infected and deceased should remind all of us to stay careful by wearing masks and avoiding large, public gatherings.

Many young people, especially in the United States, have failed to heed these warnings, attending COVID-19 parties out of carelessness instead of staying at home. In 2009, a similar situation occurred when parents held "swine flu parties" to intentionally infect their child with the H1N1 virus, and supposedly build their immunity. It's interesting to see the contrast between the reasoning behind these pandemic parties—one for fun, one for immunity—when both involve a virus not yet fully understood by scientists. You could be gambling with your life if you are not actively protecting yourself from individuals infected with a novel virus. New viruses are unpredictable and dangerous because most people experience an array of different symptoms, whether that is none or several. Unfortunately, there have been countless instances that demonstrate anyone can experience COVID-19 complications even if you perceive yourself as healthy. For everyone's protection, it is crucial that we avoid self-vaccination and wear masks to minimize the spread of infection.

So what does salt have to do with the H1N1 pandemic? During the outbreak, the CDC recommended the use of personal protective equipment (PPE) to prevent the spread of influenza in clinical settings: salt has played a significant role in developing masks and other PPE (see chapter 14). While PPE was advised for people working with individuals who had flu-like symptoms, masks were not required in public areas as it was during the

COVID-19 pandemic.

During the H1N1 pandemic, salt was valued in and out of the hospital for many reasons. In their list of recommendations to combat the cold, the CDC recommended "gargling with salt water to soothe a sore throat". A sore throat is a symptom that most of us will have experienced when we have the cold, flu or other infections. To make it clear, salt water does not protect or prevent someone from catching the cold or flu. However, a study shows that it may speed up the recovery process for the cold. While the preliminary study did not investigate whether saltwater is able to do the same with the flu, gargling with saltwater seems to be beneficial for its symptoms. After all, there must be a reason why several generations have passed down this bit of knowledge with the assumption that it's true.

Salt was also important for people who had severe cases of the H1N1 virus. Part of the treatment provided to these patients was administering normal saline (combination of salt and water at a concentration similar to our bodily fluids) through an IV. An IV is the thin tube that directs fluids (or medications) into a vein on your hand or forearm. They're typically used when drinking fluids is an unfavourable option for the patient, such as someone who is nauseous and/or requires medication that cannot be administered orally. You might be wondering then why saline would be given to patients with the flu. When someone is sick, we tell them to drink lots of fluids, such as tea or warm water. Intravenous normal saline replenishes and hydrates the body in a similar manner, which helps flu patients feel better.

In recent years, there has been a surge in "drip bars" especially in the United States that provide IV fluids without a doctor's supervision. These drips, made popular by celebrities like Kim Kardashian and Rihanna, have added vitamins and/or nutrients that are supposedly tailored to the individual's needs or desires. For example, they have been promoted as simple hangover remedies, jet lag solutions, or a quick way to get your skin glowing. If that's true, it sounds like everyone could use one of these IV drips!

Despite their appeal, it's important to know that it comes with its risks. As Dr. Shmerling points out on Harvard Health Publishing, the procedure is invasive and it is possible for the IV site to bruise or become infected. Given that these treatments are provided outside of a clinical setting, they are also quite expensive at up to several hundred dollars. So although they may make people feel great, there is no evidence that these formulas are better than what our ancestors have done for thousands of years: drinking fluids ourselves.

Now as you know, one of the best ways to combat a virus is to formulate a vaccine. As of 2020, there are several FDA-approved vaccines designed to protect humans against strains of influenza A (H1N1). They are all monovalent vaccines because they are able to immunize people against one strain of the H1N1 virus, compared to polyvalent vaccines that can protect against several strains. Of relevance to this chapter is the injectable vaccine developed by CSL Limited. Given in 0.5 mL doses, this vaccine is safe for people 6 months and older to vaccinate against the 2009 H1N1 virus. It also contains 4.1 mg of sodium chloride, which, as we know, is salt!

The H1N1 pandemic made it clear that more research was needed to better prepare the world for a pandemic. As future chapters will discuss in detail, extensive research has been conducted to improve the protective capacity of masks. One group has made a salt formulation that has been approved for funding and development to better protect our healthcare professionals from viruses such as H1N1 and COVID-19 (see Coating in chapter 12). Research has revealed that when applied to surgical masks, the salt solution can deactivate influenza viruses within a matter of 5 minutes, which has massive implications for both clinical and public settings. The coating would provide a layer of protection from germs and reduce the risk of virus transmission from touching our masks.

Masks are a vital part of PPE that symbolize the potential danger that healthcare providers face as they risk their lives to save ours. An unfortunate consequence of their work is that they are the first to identify patients who may be infected with an unknown strain of the flu or coronavirus. Many doctors, nurses and healthcare workers have passed away during pandemics due to complications of life-threatening viruses. This outcome shows that we still lack sufficient protective measures that keep our healthcare workers safe. But there is hope because of the many researchers who are devoted to developing better protective equipment that will provide a safer environment for everyone.

COVID-19 context

The COVID-19 pandemic is the defining global health crisis of the 21st century, with millions of people worldwide facing its adverse effects. From a decrease in global tourism revenue to some of the highest unemployment rates we have witnessed since the mid 1900's, this virus has quite literally taken the world by a storm. However, our strength as a global community comes in numbers, as millions of front-line workers worldwide continue to provide their services to consumers. Healthcare workers have worked tirelessly around the clock for months to protect the public and contain the virus, even as they face the highest risk for transmission and getting infected themselves. Their unwavering dedication and commitment to their jobs are commendable and should be appreciated and respected by all.

From a research perspective, scientists and researchers from various backgrounds around the globe are uniting to discover vaccines and other preventative methods that can be used by the general public to protect ourselves against COVID-19. With each research study paving the way forward for a potential cure, scientists are discovering new solutions and treatments for this virus with the hope that it will soon be eradicated or at most, contained.

The University of Edinburgh is currently conducting a study that explores the role of salt in COVID-19 treatment and cures. A pilot trial (referred to as ELVIS - Edinburgh and Lothians Viral Intervention Study) completed prior to the approved study indicated that a hypertonic saline solution used as for nasal irrigation and gargling "reduces the duration of coronavirus upper respiratory tract infection by an average of two-and-a-half days". The trial revealed that epithelial cells are able to produce an antiviral effect with the production of hypochlorous acid (an active ingredient in bleach) and chloride ions that are found within salt. These two products naturally occur within epithelial cells to clear viral infections. With the uncertainty surrounding the coronavirus and its mechanisms, researchers are unsure whether or not nasal irrigation and salt water gargling will have the same effect as seen in other strains of common colds and flus. However, it could provide a better insight into what treatment methods alleviate symptoms of COVID-19.

This study is conducted online and requires participants to complete a questionnaire to ensure they fit the criterion. Participants must also electronically sign a consent form however, they are free to withdraw at any time without reason. Following this step, eligible participants will be assigned to one of two groups, the experimental group or the control group. This study is a double-blind randomized control trial, which is considered to be the golden standard for experiments, where neither the researchers nor the participants can anticipate what group they are assigned. Based on the randomly assigned groupings, emails are sent with further information detailing the group and next steps. If assigned to the experimental or intervention group, participants will be asked to perform the nasal irrigation and salt water gargling up to 12 times a day for 14 consecutive days, or until symptoms are alleviated. If assigned to the control group, participants will be asked to continue with their normal healthcare routine and to not perform the nasal irrigation or salt water gargling. To keep track of symptoms and interventions, both groups will complete an online diary and questionnaire for 14 days every morning or until symptoms are alleviated to document any changes. Following the 14 day intervention or until well, a follow-up questionnaire will ask participants about their home environment, if other members in the household exhibited COVID-19 symptoms, and more. As an added precaution, participants are advised to abide by government recommendations and guidelines on hygiene and social distancing.

On the other side of the globe, Canadian researchers at the University of Alberta are in the process of developing a salt-coated mask designed to prevent the spread of pathogens and microorganisms. Its purpose is to increase the effectiveness of regular surgical masks by adding an additional layer of protection against airborne viruses. Common surgical masks may provide a small layer of protection however, Dr. Hyo-Jick Choi, a chemical and materials engineering professor at the University of Alberta, states how improper handling of these masks can easily increase the risk of transmission. According to Dr. Choi, "neither mask nor respirator is capable of killing a virus; which means, once contaminated, viruses can live on the surface of the filter for up to a week".

Due to the complex and unknown nature of the coronavirus, innovations such as salt-coated masks pave the way forward towards potential cures, and simultaneously eliminate treatment options that do not reduce, alleviate, or prevent coronavirus symptoms and diagnoses.

The salt coating aims to kill any viral particles that could otherwise survive on common surgical masks for up to a week. Dr. Choi explains when the virus-carrying water droplets fall on the surface of the mask, the salt will dissolve and the water will begin to evaporate. During the evaporation process, salt crystals will begin to form and destroy the virus.

Trials for a salt-coated mask have been ongoing since early 2015 however, Dr. Choi and his team's efforts are more important now than ever. His team has tested this coating on three different influenza virus strains, all of which became inactive within a 30 minute timespan of being exposed to the salt surface. It is their hope that this product will soon be manufactured with an expected release date between 12 to 18 months.

In Jamaica, a country with less than half of the government funding and research facilities than North America, one of their top medical practitioners, Dr. Alfred Dawes is also suggesting something similar to that of salt-filtered masks. Dr. Dawes and other medical colleagues have been experimenting with table salt and everyday absorbent materials that could be used to develop a disposable filter. Materials included coffee filters, napkins and toilet paper that were soaked in a hot salt water solution, dried, and within 30 minutes, results show how it could effectively kill any virus that comes into contact with it. The salt crystals reform during drying and the sharp edges are able to cut through the outer coating of the virus on contact, giving them the ability to kill the virus within 30 minutes.

"I know the salt filters need to be evaluated as to how effective they are, but we do not have that capability in Jamaica. Neither do we have the time to wait. I hope someone with the resources will be able to take this idea and validate it in the field. But I strongly believe in the science behind it and I know it will cause zero harm to use the filters as long as we continue hygienic practices and social distancing," said Dr. Dawes in a recent interview. The process of making these filters are simple, cheap, and hopefully effective if worn on the inside of cloth masks. He is strongly recommending the use of the salt filter and hopes it will help decrease the transmission of COVID-19.

The University of Alberta is also partnering with a startup company to prevent the spread of viruses and viral infections through high-touch surfaces, such as door levers and handrails. Outbreaker Solutions, an Edmonton based startup that have patented a "self-sanitizing, antimicrobial surface based of compressed sodium chloride (CSC)". This durable surface can be installed on anything that is frequently touched by hands or other objects, and kills microbes in minutes whereas other commonly used surfaces retain them hours or days.

Brayden Whitlock, Outbreaker's director of research and PhD candidate in the U of A's Department of Physiology stated how "the public is starting to catch up with long-known statistics showing that the vast majority of infections are spread by our hands, rather than through the air. This will become a more important field as they learn more about infection control, and it's being forcibly accelerated by our current pandemic".

Outbreaker Solutions have partnered with Xiao-Li Lilly Pang, a professor in the Department of Laboratory Medicine and Pathology, and program leader of microbiology at the Provincial Public Health Laboratory.

Combining forces with Pang's expertise and her team of researchers focused on virology, and Outbreaker Solution's product and monetary support as a part of UAlbertal Health Accelerator, both parties are convinced that their innovation could drastically alter and potentially decrease transmission rates between individuals for all viruses, especially COVID-19.

The overall purpose of this project is to provide an additional layer of protection against infections occurring in the first place, rather than transmission between people and objects, with the emphasis being on health-care workers and patients. This proposed idea would not force people to change their daily behaviour (as various studies have had underwhelming results when attempting to alter an individual's habits) and requires no extra maintenance or cleaning. Furthermore, there are no approved drugs or vaccinations that prevent the transmission of coronavirus, nor are there sanitizers or disinfectants that have the ability to inactivate the virus. If this product has a virucidal effect, it would be an ideal and effective material to coat layers on hard surfaces in high-touch facilities including (but not limited to) hospitals, schools, long-term care centers, and more. While this study requires more funding and time to advance their research, the goal is to take their innovation to public settings as soon as possible and cater to larger markets and products such as hospital bedrails and public transit handrails.

Salt's multifaceted roles as an essential cooking ingredient, antibacterial agent, and mask coating has continuously shown to be beneficial and favourable in a variety of settings and situations. However, its versatility proves to be more important now than ever with COVID-19 cases continuing to rise globally. Alongside medical practitioners, researchers, and scientists, it is our hope that these studies on salt's effectiveness as a mechanism to protect ourselves proves to be vital in the race to find a cure against the coronavirus.

Saline in Masks

Since the start of the COVID-19 pandemic, people have been trying various DIY (do-it-yourself) home remedies to prevent contracting the virus. While many of these methods can prove to be harmless, there are a number that could cause adverse effects. In May of 2020, the Center for Disease Control (CDC) in the United States of America conducted a study of 502 individuals and determined that 39% of the sampled population had misused disinfectants in an attempt to kill COVID-19. Some of these high risk behaviours included using bleach on food products (19%), misting disinfectant spray (18%), inhaling the vapour from cleaning products (6%) and even drinking or gargling various forms of bleach or disinfectant products (4%).99 A quarter of these respondents had detrimental side effects as a result of their risky practices.99

However, there is one DIY method that people have been using since the first SARS outbreak that may prove effective to combat COVID-19, but not in the way you might think. During the SARS outbreak, many turned to traditional Chinese medicinal practices to combat the virus, some of which included consuming iodized salt and washing their body in salt water. While none of these methods ever proved fruitful in a laboratory setting, they may have had the right idea regarding the use of salt to prevent the contraction of viruses. A study published in 2017 by Fu-Shi Quan and others discovered that using salt in masks may actually be helpful to filter particles from harmful viruses. The researchers were investigating the efficacy of adding salt to the filtration unit of a surgical mask as a reusable and cheap method to protect citizens from outbreaks like COVID-19. The basis for the study was the thought that the sharp edges of salt crystals could cut through and destroy the viral particles trapped in the respiratory droplets of viruses like COVID-19 or destroy airborne viruses in the same manner.101 This discovery came about when one of the researchers, Dr. Hyo-Jick Choi was in the midst of developing an oral vaccine. Vaccines work by introducing your body to a small amount of a weakened or dead version of the virus it is meant to protect you against. In doing so, it provides immunity from the virus without causing severe illness and should the virus be contracted, your body will be able to recognize it and fight it off. However, Dr. Choi found that the sugar he was using as part of the oral vaccine was neutralizing the weakened form of the virus, thus rendering his vaccine unsuccessful.102

He determined that this occurred due to the crystallization of sugar and its sharp edges cutting through the virus. This fortunate stroke of serendipity is what led to his idea of utilizing salt's similar crystallization properties as a method to prevent the contraction of viruses. He decided upon the use of salt instead of sugar since the former has more uniform crystallization properties, making it easier to work with.

Dr. Choi began experimenting how different salt solutions, commonly known as saline solutions, affected different viruses. It was at this time that him and a team of other researchers began testing the efficacy of salt-coated masks. The results of the study, published in 2017, showed that the masks could successfully neutralize three strains of the Influenza virus. In a regular surgical or N-95 mask, although the mask may serve as a barrier between one's face and the viral particle carrying droplets, the mask itself does not kill the virus. This results in a chance of contact transmission, which encompasses physical contact between an infected person and another individual or when you contract the virus by touching objects carrying the virus. With Dr. Choi's salt-coated mask, the droplet quickly absorbs the salt and the liquid evaporates, leaving only the salt and the viral particle. It is at this point that the crystallized salt is able to pierce through the virus, inactivating it.106 The study showed that it took 30 minutes for the virus to be completely destroyed, although it becomes inactive within 5 minutes.106 Dr. Choi currently has a provisional patent on these masks and hopes to be able to mass produce them soon. This invention could revolutionize how we combat viruses in a cheap, effective, and accessible manner.

However, it is important for these masks to be used the right way in order to be effective. After Dr. Choi's study was published in 2017, there were reports of citizens soaking their masks in salt-water as a method of protection. In response to this Dr. Choi remarked that this is "not how the masks work" and emphasized the importance of using the masks currently available correctly. Although the invention is quite simple compared to other technological inventions, it is still not a DIY solution, as there is a specific salt solution that must be used and the salt must be embedded within the filtration unit of the mask. Until these masks begin to be produced on a commercial level, your best bet is to stick to your everyday surgical or cloth mask when in public to keep yourself and others safe. The root cause of people passing on COVID-19 despite wearing masks is typically that people will regularly move their masks around, take them off and put them back on, and adjust them frequently. This greatly increases the chance of contact transmission even if you are wearing a healthcare grade mask like the N-95. However, the main risk arises when citizens are wearing surgical or cloth masks, because unlike front line workers they are not trained in how to best prevent transmission while wearing a mask. This problem could largely be eliminated with a mask that renders the virus containing droplets ineffective in a matter of minutes.

Despite the potential that these masks have, the current pandemic has proved that many are unwilling to simply wear a mask to protect those around them. This brings about the question, how can we use these masks for a future pandemic if people are refusing to wear masks in the first place? According to a study conducted by the CDC over 17% of Americans are unwilling to ever wear a mask in public, despite it being one of the simplest ways to protect yourself and others from contracting COVID-19 and the immense scientific research demonstrating its efficacy. A review of 172 studies published in The Lancet during June of 2020 found that if people simply cover their face in public settings, it can minimize the risk of catching and transmitting COVID-19. The co-author of the review, a physician and epidemiologist at McMaster University, stated that it showed "[i]n multiple ways ... the use of masks is highly protective in health care and community settings." Face coverings work by preventing large respiratory droplets from infected individuals from spreading into the air, onto others or onto objects. It is important to note that in order to truly minimize the spread of COVID-19 individuals must wear masks in conjunction with abiding by social distancing guidelines and regularly washing their hands.

Many anti-mask proponents say that laws in place mandating masks in public settings is an infringement of their "freedom" and as a result they become morally outraged at the request to don a mask.112 David Abrams, a clinical psychologist at New York University, noted that some may also refuse to wear a mask on the basis that it makes them look vulnerable to others. Regardless of what the reasoning may be, the science is clear and it is shocking to most that people would rather stand their ground on what they perceive to be an issue of "freedom" than simply wear a mask when outside to protect others around them. This has proven that for some people, basic altruism and the concept of wearing a mask for a short time period to potentially save another's life is not enough. This begs the question, what can government officials do to keep their communities safe? The easiest way to get around people's desire to not wear masks is to make it mandatory by law to wear them in public settings when indoors. Although this could be seen by some as a natural first step amidst a pandemic that began on March 11th, 2020, the United States of America still has not made masks mandatory on the federal level as of August 2020, despite the majority of individual states doing so. Back in March, when the director-general of the Chinese Center for Disease Control and Prevention, George Fu Gao, was asked what mistakes countries were making in regards to handling the pandemic, he noted that "the big mistake in the U.S. and Europe, in my opinion, is that people aren't wearing masks. This virus is transmitted by droplets and close contact. Droplets play a very important role – you've got to wear a mask, because when you speak, there are always droplets coming out of your mouth. Many people have asymptomatic or presymptomatic infections. If they are wearing face masks, it can prevent droplets that carry the virus from escaping and infecting others." A study conducted by researchers at the Virginia Commonwealth University found that in countries where there was a cultural norm to wear masks or the government outwardly supported mask wearing from the beginning of the pandemic, mortality rates increased only 8% per week, compared to a 54% increase in the rest of the countries.

The onset of the COVID-19 pandemic and its unprecedented global spread has taught us much about ourselves and our communities, and has helped to shed light on the realities of our societal structures. Traditional medical wisdom has continued to inform and enhance innovations in modern research, as we find new ways to integrate the useful crystallization properties of salt. The pandemic has additionally emphasized the importance of community responsibility, through wearing masks or otherwise, and the need to build a culture that prioritizes the wellbeing of the whole.

Salt Filters

Everything on this planet, that we call Earth, requires water to survive. From the smallest bacteria and macrophage, to some of the largest animals alive such as elephants and blue whales. In fact, even those long gone such as dinosaurs and mammoths required water. When it comes to most mammals though, we can only survive on fresh water, not salty or brackish water. So even though 71% of the Earth's surface is covered in water, 97% of that water cannot be used by us.113 Unless, of course we find a way to desalinate and cleanse the water. For those of us who live in 1st world countries, water might not seem like such a commodity, but thousands of people all around the globe struggle to find the most basic source of clean and fresh water. Living in Canada, we have been sheltered, because Canada has about 20% of the world's freshwater resources for only 0.5% of the world's total population. How does all of this relate back to salt, is the question? Well, most of the world's water supply is "salty", aka has a high concentration of salt, and thus not drinkable by humans. The goal for technology is to then be able to provide freshwater from salty water, thus meeting the world's continuous growing demand for freshwater.

There are several ways we can desalinate water, the two main ones being distillation and reverse osmosis. Distillation is the most basic process of purifying a liquid either by heating or cooling. Specifically when it comes to salt, boiling is used since water has a significantly lower boiling point than salt. So the process for this involves bringing saltwater to a boil, which will cause all the water particles to evaporate leaving the salt crystals behind. The evaporated water particles are then collected and condensed, thus providing you with a pot of fresh drinkable water.113 Although a great idea on the small scale, this possesses various problems when industrialized. In fact the biggest barrier to mass producing fresh water using distillation is the energy costs. Due to the large amounts of energy that go into bringing the water to a boil, and then also condensing the water this is not the preferred form desalination for most large corporations.

Instead, a process called reverse osmosis is used. This process was actually discovered in 1969 by an engineering student named Dean Spatz, and now today it is the most popular method of desalination available1. In fact, there are over 18,000 desalination plants that use reverse osmosis in over 150 different countries, with over 300 million people relying on it for fresh, clean water. The process of reverse osmosis itself is quite lengthy and will be examined next115.

In reverse osmosis, the key component is a special semi-permeable membrane (a salt filter per say) that helps separate not only salt from water, but any kind of biotic materials present as well. Seawater is actually full of living and dead organic and inorganic molecules. This includes phytoplankton, zooplankton and bacteria which exist in the millions115. Upon closer examinations one would also find marine plants, carbonate particles and fish and invertebrate mucus. Most of this material is actually held suspended in water, and thus the first step of reverse osmosis is removing all the particles in the suspended load of seawater. The seawater is usually treated with some kind of a coagulant which causes the suspension load to fall to the bottom of a settling chamber, with the cleaner water above it115. The next step is to pass the water through a multimedia filter. This again allows more suspended material, and some larger abiotic particles to be filtered out. The multimedia filter is a filter formed from coarse gravel, sand, garnet and anthracite usually107. The seawater is passed through this with a pretty slow flow rate, such that a lot of particles can be caught, and the water cleaner leaves the chamber. This whole process was the pre-cleaning/pre-treatment component. The next step is when the reverse osmosis filters actually come into play. Osmosis is the process in which water will move across a semipermeable membrane to the side that has a higher concentration of solute. For example, in calls, if the concentration of salt is higher outside of the cell than inside, water particles will diffuse out of the cell into the surroundings . Thus, in reverse osmosis (RO) filters, we are literally reversing osmosis, by having water move from the higher concentration solution to that of the lower concentration solution. This is achieved by pushing the water through the membrane at high pressures, at about 800psi117. This high pressure, pushing H2O particles through the membrane, while most of the salt and mineral particles are left behind, thus producing fresh water. Most desalination plants will have multiple spiral-wound RO's lined up one after another. As the water continues to pass through each membrane, the concentration of salt and minerals is significantly lowered, with the water after the last membrane considered freshwater117. This water, though considered fresh, is not drinkable yet. The last step in this process is using UV light in order to kill any bacteria, algae, and fungi growing in the water. The UV light will cause DNA damage and thus lead to apoptosis, cell death, within the microorganisms117. After all this, the water is now clean and drinkable for those far and near.

Latest technology in salt filters for seawater, actually includes using graphene filters. Although reverse osmosis is more cost and energy efficient than distillation, it is inefficient and ineffective in remote areas and areas farther from major cities and sea coasts. Enter recent advances in technology, and the development of hybrid graphene oxide membranes, that can be used to filter salt from seawater cheaply and effectively . The challenge with filtering out salt ions from water, is the fact that salt ions are actually smaller than water ions. In RO, the high pressure is what pushes the freshwater through, but that is also what makes it expensive.

Graphene is a thin, two dimensional membrane composed of carbon atoms that are bonded together in a hexagonal pattern. Graphene is highly hydrophobic and thus cannot be used as a membrane. However, graphene oxide is an oxidized form of graphene that is more hydrophilic due to the oxygen molecules present. This membrane is highly selective for water molecules, but impermeable (to a certain extent) to salts, minerals and other impurities. The real problem with this filter is being able to build a membrane that can last for long periods of time, months if not years . In particular, chlorine is extremely harmful to these membranes, quickly breaking down the oxide-carbon hybrid structure. As such, recent research is focused on incorporating pure graphene into the membranes and extending membrane life. Such filters are in the trial phases and have shown to block 85% of salt molecules from passing through, which although really good needs to be improved on118! A new age of salt filters is being ushered in with the development of new membranes and nanotechnology118. Which is great news for 20% of the world populations whose basic fresh, clean and drinkable water needs are still not being met118. Hopefully in time and with technology, a new, cheap and permanent solution will be brought to light.

Salt filters, although relevant to water desalination can also be referring to a completely different type of technology known as a salt chlorine generator. Salt chlorine generators are used in swimming pools and hot tubs as a sanitation method. This process is dependent on the basic chemical principle of electrolysis . Electrolysis is a process in which electric current is passed through a substance to cause a chemical reaction118. The process takes place in an electrolytic cell, a setup that has a negative electrode and a positive electrode. In this process you have two main components, the cathode (positive electrode) and the anode (negative electrode). The anode is where oxidation takes place, the loss of an electron or gain of an oxygen molecule . The cathode is where reduction takes place, the gain of an electron or loss of an oxygen. Both the electrodes are dipped in solutions containing either positive or negative ions, and are connected by a salt bridge. The salt bridge acts as a connector and also helps maintain the cell's neutrality in the electrochemical reaction. An electric current is passed through this apparatus, starting at the cathode (cause reduction) flowing through the anode and then back around forming a closed circuit120. When this process is used with a low voltage and salt water, also known as sodium chloride and H2O, the products of the reactions yield hydrogen gas and hypochlorous acid. The hydrogen gas bubbles out of the pool water into the atmosphere while the hypochlorous acid acts as a disinfectant121. The acid will leave the electrolytic cell and continue disinfecting the pool, until eventually it will recombine with water and salt molecules present in the pool to reform salt (NaCl) . This process is a circular loop, repeating itself, which is why it is so popular amongst pool and hot tub owners. It is also quite cheap, and relatively energy efficient121.

Salt filters can be referring to a variety of objects, from simple distillation and reverse osmosis to graphene oxide filters to salt chlorine generators, they are all extremely useful and handy. Millions of people every day don't have access to direct sources of clean, fresh drinkable water, and reverse osmosis makes that possible . It allows us to remove any abiotic and biotic components from salty and brackish water, including salt itself which can be impossibly hard to remove on an industrial scale122. Moreover, 97% of the earth is actually covered in seawater, so being able to access this untapped natural resource makes access to water easy for everyone, not just the top percentage of the world122. Recent advances in salt filters take such goals a step further, wanting to reach the most rural communities, and provide energy efficient fresh water to millions of people in a large scale way. Furthermore, salt filters also have an impact on the lives of people every day. Salt chlorine generators are a common salt filter used in swimming pools and hot tubs, which help disinfect water supplies and keep it clean and possible to swim in. Without salt filters clean pool water and clean drinkable water would not be possible today121.

DIY solutions

While most of us believe that COVID-19 is an entirely new virus altogether, according to the CDC, various strains of the coronavirus were first identified in the mid-1960's. Coronaviruses are named due to the crown-like spikes on the virus' surface and are a group of related RNA viruses with the ability to cause a wide range of diseases and infections in mammals and birds. From the common cold to more lethal varieties such as SARS and COVID-19 (a novel coronavirus), all strains cause respiratory tract infections with no known cures and vaccinations as of today. There are four main sub-grouping of coronaviruses; alpha, beta, gamma, and delta. These are also the most common sub-groupings of coronaviruses that people worldwide get infected with. However, some of these coronaviruses can evolve and become a new human coronavirus, such as SARS-CoV, MERS-CoV, and 2019-nCOV (also known as SARS-CoV-2 or COVID-19).

With COVID-19 cases continuing to rapidly surge across the globe in countries such as the United States and India, citizens have begun to take matters into their own hands and implement DIY solutions into their daily routines. While some of these practices may seem absurd, some are convinced that these methods not only alleviate, but also minimize the likelihood of contracting the virus.

These so-called "immune boosters" are gaining steam amongst the general public, even though various accredited institutions and companies have spoken out and strongly condemned such practices as it poses a severe threat to consumers and their health. To make matters worse, various political figures, including Donald Trump, have outrightly spoken about injecting bleach into the bloodstream or irradiating patients' bodies with UV light during televised broadcasts and interviews. These falsified claims were immediately lambasted by medical communities worldwide, especially since disinfectants are hazardous substances and can be poisonous if ingested. Even external exposure can be dangerous to the skin, the eyes, and the respiratory system. While he may be a widely renowned political figure, he is not a medical professional who has the ability to give such advice and as a result, his ideas could have detrimental and fatal results for all those that follow. While Trump posed the idea of injecting disinfectants and UV light rays as a question, statements such as these, made by political figures on live broadcasts, can be easily taken out of context. During an interview with NBC News, pulmonologist Dr. Vin Gupta said "this notion of injecting or ingesting any type of cleansing product into the body is irresponsible and it's dangerous. Another pulmonologist at Zuckerberg San Francisco General Hospital "warned that even breathing fumes from bleach could cause severe health problems [...] it's a totally ridiculous concept". Once again, this proves the importance behind using credible sources to stay updated on the latest trials of preventative measures and vaccines for COVID-19. Moreover, it is crucial that the public allow researchers to do their job and guide us on ways to prevent the spread rather than us becoming self-proclaimed medical professionals.

Bleach and its antimicrobial properties have the ability to kill harmful microorganisms due to the active ingredient, sodium hypochlorite. It is effective in killing bacteria, fungi, and viruses however, it is extremely unsafe and hazardous to ingest, inhale, or inject into the body. An alternative to bleach that is not directly harmful to the body involves the use of salt and walter, also referred to as a saline solution. As of today, it is not proven that saline solutions can kill the coronavirus however, saline sprays have been thought to help relieve nasal congestion and sore throats; the most common symptoms of colds and flus. Dr. Sandeep Ramalingan and his medical colleagues conducted a study in 2019 that studied 60 people who had developed a common cold. 25% of the colds that these people developed were confirmed cases of a common coronavirus with mild symptoms of nasal congestion, sore throats and a low-grade fever. 30 people were assigned to use a salt-water gargle and nasal inhalation, and the other half acted as the control group in the experiment with no interventions required.

The results showed that the "hypertonic saline uses had milder symptoms and the duration of the cold was significantly blunted". Moreover, saline sprays significantly reduced the rate of transmission upwards of 35%; an important rate and finding in the context of researching for a cure or vaccine of the novel coronavirus. A somewhat similar study examined treatment options for 400 children, aged 6-10 years old, that developed colds and the influenza virus. Over the course of 12 weeks, approximately one-third of the children were treated with conventional over-the-counter medications, while the others were prescribed nasal saline sprays. Following this intervention, the results established that those who used the nasal saline spray on a regular basis had quicker resolution of infection and less reappearance of the illness.

Another common DIY 'solution' that has been circulated for decades, even before today's novel coronavirus, is using the same saline solution for gargling. Advocates for saltwater gargling say this method can eliminate the virus and prevent it from reaching the lungs via the throat. The World Health Organization's (WHO) website offers an immense amount of information surrounding COVID-19 and addresses these misconceptions regarding saltwater gargling and rinsing the nasal passages with saline. Unfortunately, "there is no evidence that regularly rinsing the nose with saline has protected people from infection with the new coronavirus. There is some limited evidence that regularly rinsing the nose with saline can help people recover more quickly from the common cold. However, regularly rinsing the nose has not been shown to prevent respiratory infections". Again, while saline solutions have shown to be effective in reducing transmission and alleviating symptoms or duration of the virus, it has not been proven to be the sole mechanism to alleviate all symptoms and guarantee protection against the novel coronavirus of today.

More research into the mechanisms of saline solutions found that if the salt and water mixture is combined with an electrical current, it is capable of making hypochlorous acid, HOCl. This weak acid has a pH similar to that of a citrus-based juice and is a non-toxic substance that can be ingested, unlike sodium hypochlorite, the major constituent of bleach. HOCl is made naturally by our white blood cells used in healing processes and protection against invading bacteria, fungi, and viruses. It is also the "active ingredient in electrolyzed water, which is an industrial technology used for green cleaning and sanitizing". Electrolyzed water is the product of salt, water, vinegar, and an electrical current to change its overall chemical structure. It is as effective as bleach but with the added benefit of having no harmful chemicals, fumes, pesticides, or toxins. With the correct proportions of salt, water, and vinegar, the electric current forces the molecules of these constituents to break apart and form two new molecules; hypochlorous acid (HOCl) and sodium hydroxide (NaOH). The hypochlorous acid is considered to be "nature's super powerful disinfectant" and is the key component that gives bleach its well-known antimicrobial power. Due to vinegar's acidic nature, when the pH of the solution is lowered to its optimal level, HOCl is produced. This naturally occurring substance in our body is considered to be our immune system's first response against invading microorganisms, alongside our white blood cells. The other substance produced is sodium hydroxide (NaOH), which is a combination of a salt and a hydroxide ion that is composed of oxygen and hydrogen (OH-). NaOH is a common detergent used in various household items including toothpaste and cleaning products.

Hypochlorous acid differs from bleach in more ways than one. Firstly, they have different chemical formulas. The formula for sodium hypochlorite (bleach) is NaOCl and the formula for hypochlorous acid is HOCl. From a chemist's perspective, HOCl has no charge and a relatively low molecular weight. This allows it to penetrate cell walls more easily compared to other disinfectants, proving its efficacy. Conversely, the hypochlorite ion found in bleach is unable to penetrate and diffuse through cell walls, and its negative charge causes it to become electrostatically repelled from these same cell walls. NaOCl is a larger molecule in size and is also the reason for its slow diffusion rate when compared to HOCl. Bleach is defined by three properties; its pH of 11+, being at a concentration high enough to remove colours from most fabrics and clothing, and being composed of 99% NaOCl. When the pH of bleach is lowered, allowing the solution to become more acidic, the hypochlorite converts into hypochlorous. However, the most fascinating point is that more bleach would be needed to achieve similar or the same microbial properties and power as hypochlorous acid, which explains why HOCl is used in a wide array of applications. It is "suggested that HOCl is 80 to 120 times more efficacious than sodium hypochlorite".

While it may not be known to the general public, hypochlorous acid has several uses in the healthcare and medical field in wound care and healing processes. Its efficacy in killing bacteria without harm to our body has been extensively researched in various trials and studies. This poses the question - why is bleach so widely advertised as a disinfectant when there is a safer alternative available? The answer to that million dollar question is money. Even though electrolyzed water was initially discovered decades ago, the equipment and technology used to produce it can cost upwards of $10,000. Additionally, hypochlorous acid dissipates over time, which is why bleach has a longer shelf life and a larger appeal to companies and consumers. Scientists and researchers are optimistic that advances in technology will help electrolyzed water and hypochlorous acid become more readily available and desirable to consumers as an alternative to bleach.

While DIY solutions may be picking up steam across the Internet and amongst people worldwide, the largest takeaway from this chapter must be that fact-checking these DIY practices is vital for our safety and health. It is not in our best interest to rely solely on the words of high-profile politicians and public figures with no medical background. As responsible citizens, we must do our part in protecting ourselves and others during a pandemic by following the guidelines and recommendations outlined by accredited institutions and medical professionals. With their expertise and knowledge, it is our hope that a treatment will be available soon for those diagnosed with COVID-19 and for those in recovery.

Infusion and coating

So far, you've learned about how salt has been used in various applications within and outside of healthcare. From sidewalks to french fries to masks, salt has incredible versatility. Some of its applications include two processes called infusion and coating, and research has shown the benefits they may have for our health. Salt infusion can help alleviate symptoms of respiratory conditions, while researchers have developed a salt coating that may provide us additional protection from life-threatening viruses.

Infusion

Imagine you're at a beach, sitting on the warm sand by the sea. You look out, and you look up at the sky as you relax and breathe in the air around you. A day spent by the shore leaves all of us feeling refreshed, relaxed, and overall in a better mood. Salt therapy provides a similarly pleasant experience, but instead of sand on the ground, there's salt. In fact, those constantly on tight and busy schedules will be delighted to know that a 45-minute session of salt therapy is approximately "equal to three days by the sea". Three days! That sounds like a good trade-off, wouldn't you agree?

The therapeutic benefits of saltwater have been recognized since the Ancient civilizations, but the first formal record of salt therapy was in Poland. A doctor noticed that miners who worked in salt caves developed fewer respiratory diseases than other miners (who, for example, dug for ore or coal). Respiratory diseases encompass the conditions that affect our ability to breathe: the common cold, chronic bronchitis, allergies, asthma are just a few. After the Polish doctor's discovery, people sought out to find salt caves to cure their own conditions, and this was termed speleotherapy. Although there are few salt caves here in North America, there are many spa and wellness centers that now mimic the experience through an artificial process.

Salt has several properties and functions that make it a special mineral. Our body needs it to perform metabolic processes and maintain our energy as we go about our day, but it also has antibacterial and antimicrobial properties. The salt on the floors and walls of salt caves and man-made spas are made of imported sea salt that comes from the ocean or pink Himalayan salt mined from Pakistan. Both have many trace minerals (with pink Himalayan salt having slightly more) that contribute to their moisturizing and anti-inflammatory properties, which are both beneficial to our skin and lungs. But don't be mistaken: both types of salt are still mainly composed of sodium chloride, so when they are ingested, they don't provide notable health benefits compared to typical table salt.

Salt therapy is scientifically named halotherapy, where halos means salt in Greek. It refers specifically to the inhalation of miniscule salt particles; this is done through a halogenerator that sprays salt-infused air in designated intervals. Halogenerators are used in active salt rooms, whereas passive salt rooms have walls, floors, and fixtures made of salt but do not spray it. The key to benefitting from salt therapy comes from inhaling these tiny molecules of salt, where it will reach your nose, throat, and lungs. As a result, passive salt rooms are not known to improve our health, but they do provide a space for relaxation that can be quite beneficial. Both types of salt rooms are kept temperature-controlled at 21-24 degrees Celcius to maintain a relatively dry environment and maximize the salt's benefits.

Inhaling salt in these salt rooms found throughout Europe and America is recommended as a natural treatment for many health concerns. They can improve skin conditions like acne and eczema, acting as a detox for the body. The following lists some, but not all of the potential health benefits that salt therapy may have for physical health:

- Respiratory conditions

o   Common cold
o   Asthma
o   Allergies
o   Bronchitis
o   Cystic fibrosis
o   Sinus headaches
o   Chronic Obstructive Pulmonary Disease (COPD)

- Skin ailments

o   Eczema
o   Psoriasis
o   Rashes
o   Acne

Salt therapy is also reported to improve mood, energy, and sleep patterns while relieving stress. The calm and soothing environment that spas are known for most likely contribute and add to the therapy's positive effects.

So why is inhaling salt beneficial? A Senior Scientific Advisor of the American Lung Association explains that salt therapy could thin the mucus in the airway for those with respiratory conditions, which makes it easier to cough it out. Being able to cough the mucus out would help individuals alleviate their symptoms, even if the effects are temporary. It may also reduce inflammation in the lungs which explains how salt may improve the respiratory diseases mentioned above. A study that recruited over 150 patients found that salt therapy improved breathing in individuals with cystic fibrosis, a genetic condition that causes mucus to build up in the lungs and makes breathing difficult. Asthmatic individuals have reported that they could breathe easier following 2 weeks of salt therapy as an additional therapy to steroid inhalers. Similar results were also found in pediatric patients. While these results are promising, studies like the latter two recruit a small sample of individuals that make it difficult to generalize to the population with these conditions. Small participant size is often an issue for clinical studies, as they require participants who meet specific criteria/diagnoses and also consent to subjecting themselves to therapy or medication.

Moreover, salt therapy has its downsides too. Due to the warm environment of salt rooms, some may develop skin irritation that disappears once the treatment session has ended. But for those of you who have sensitive skin and are prone to irritation, keep that in mind before an appointment. Furthermore, since the treatment seems to loosen the mucus in the lungs (as we discussed in the previous paragraph), it can cause coughing for those with respiratory conditions. If you have an existing respiratory disease, it is especially important that you consult with your doctor and discuss the therapy before booking yourself for a session. Doctors do not strongly advocate for salt therapy as there is still insufficient evidence that strongly supports its benefits. Many of the studies that exist are not considered methodically valid in the eyes of researchers. In the future, we need more studies conducted on various respiratory/skin conditions with larger participant sizes for physicians to truly know whether the therapy is more helpful than harmful.

With this all in mind, we aren't discouraging you to try salt therapy. There's extensive anecdotal evidence that supports it as an enjoyable, effective, and comparably inexpensive treatment option for several conditions. Not to mention, the atmosphere of the environment itself provides time to relax or meditate, which can have a positive impact on our mental health as well.

In the future, we hope that research demonstrates salt therapy's benefits so that it can be recommended for those who would benefit from it.

Coating

Scientists have been incredibly innovative and resourceful during the pandemic. One area of interest has been finding ways to improve the amount of protection that masks provide from viruses. Experts predict they will remain a mainstay for, at the very least, the next year. Masks play a significant role in healthcare but also in public places where interaction with others, even with distancing measures, is bound to happen.

One critical thing to understand about surgical masks is that they were not designed to destroy viruses or prevent their spread: they were meant to act as a barrier that protects against 8infectious droplets in clinical settings. Think of the filter in surgical masks like a sieve, where big particles are unable to pass through it. However, respiratory illnesses such as COVID-19 are made up of smaller droplets and aerosols (which are tiny particles that float in the air) that can pass through the sieve, or mask. The difference between surgical masks and N95 respirators lies there: N95 respirators act as a finer sieve, so they are better at filtering both large and small particles out, but they're harder to breathe in. N95 respirators are also quite costly and require an evaluation to ensure that it fits perfectly on the individual. During times of shortage such as a pandemic, N95 respirators are only meant for use by healthcare providers who have close contact with infected patients.

The main issue with current masks that are manufactured is that they do not deactivate viruses. By lacking this function, we put ourselves at risk for transmitting the virus to other surfaces and ourselves whenever we touch our mask. Assistant Professor Hyo-Jick Choi and PhD student Ilaria Rubino have worked with their colleagues at the University of Alberta since 2015 to study surgical masks. They focused on surgical masks for good reason: they are commonly worn by healthcare professionals, and favoured by the public. Yet, as we've established, these masks aren't that effective with viruses. So the researchers came up with a salt solution—primarily made of two salts, sodium chloride and potassium chloride—that can be applied on surgical mask filters to attack viruses.

The chemistry behind the coating comes from a common problem Choi encountered with formulating oral vaccines. When liquid vaccines undergo the process of turning solid, sugar crystals (meant to stabilize the vaccine) form and deactivate the virus, rendering it inactive and unable to provide immunity. Choi took this phenomenon and applied it to salt filters. As aerosol water droplets of the virus touch the salt-coated filter, the droplet dries while it absorbs salt. Increased salt absorption causes the formation of salt crystals, and with their sharp edges, the crystals pierce the virus and destroy it. You can think of the process being similar to "a sharp, briny needle piercing a balloon".

Research on this salt formulation has been conducted in both Canada and South Korea, showing promise for its ability to deactivate three strains of the H1N1 influenza viruses on a contaminated mask within 5 minutes. Within 30 minutes, the viruses were completely destroyed! With funds provided by a not-for-profit organization that Rubino works with, Mitas, and the National Sciences and Engineer Research Council (NSERC), Choi hopes that they will get the product on the market in 2021.

Research is showing a viable path to a future of better masks that will protect our healthcare providers and the public. But even with these improvements, it's important to not lose sight of how we can minimize our own risk and others. The following are recommendations on how to wear a mask (but we hope you are already following them):

- Ensure that your mask fits your face as close as possible, with the nose covered and no side gaps
- Do not lower your mask to your chin
- Wash your hands before and after touching your mask
- Avoid touching mask filters
- Do not reuse or store used masks
- Never put your used masks in or on potentially contaminated places (i.e. pockets, countertops)

Door Levers and Handrails

Every day, at home or work or school, we are surrounded by germs—tiny microorganisms that can make us sick. Germs encompass bacteria, viruses, fungi, and one-celled organisms, and they live on everything that we touch: handrails, door knobs, office desks, computers (especially keyboards), cell phones, light switches, remote controls, elevators. If we consider handrails (i.e. on the bus, subway, escalator) alone, numerous individuals will have already touched them by the time it's your turn, and likely before they've been disinfected. One might say it's similar to shaking hands with hundreds of people!

So far we've talked about how salt may be effective in killing influenza viruses, and its therapeutic benefits for our health. It has also played a key role in research that aims to improve the protection of healthcare workers. Yet scientists have found another use for salt that minimizes virus transmission in public places. Aren't they incredible?

As a consequence of the pandemic, there has been overall higher demand for cleaning/disinfectant products. As cases of coronavirus began to rise, Canadians saw an unprecedented situation unfold: the toilet paper shortage. While toilet paper was the item of interest to newspapers and the public, antibacterial wipes, hand sanitizer, and other cleaning products flew off the shelves in grocery stores everywhere. Many people stocked up on non-perishable food with the high possibility that self-isolation would become enforced for an indefinite amount of time.

Then the public began to raise an important question: was it possible for the new coronavirus to be transmitted through surfaces? What risk do grocery bags and groceries themselves pose for our health as we unpack them indoors? The answer to these questions have become more evident through ongoing research. Thus far, there have not been coronavirus cases directly linked to surface-to-person transmission. However, one thing is clear: viruses live much longer on surfaces than we think.

The flu virus can survive on hard surfaces, such as desks or keyboards, up to 48 hours. Now consider the number of times we are touching them and switching to another task (i.e. going to the bathroom, drinking coffee, writing in a journal) before we disinfect these items. It's not hard to see how viral particles can easily spread to other people, and other objects, in a matter of two days. The following factors alter how long the virus will survive on a surface:

- Type of surface
- Amount of virus
- Temperature
- Humidity of the environment

Despite the limited time and resources available to scientists, several studies have shed light on how long SARS-CoV-2 (the virus that causes COVID-19 disease) can last on certain surfaces. A study by Doremalen and colleagues that is supported by the National Institute of Health (NIH) revealed that the virus can survive on a surface from a couple hours to several days. In fact, their results suggested that it has a similar survival rate to the virus that caused the SARS outbreak in 2003. If viral droplets are small enough to linger in the air, they can remain afloat for several hours. This alone explains why it is crucial to minimize public outings, wear masks near others, and avoid touching our faces to lower the risk of transmission.

The following table "How long coronavirus lasts on surfaces" outlines the results of the NIH study. Plastic surfaces can keep the virus alive up to 3 days, but surprisingly, so does stainless steel. Given how many household and public items are made of these materials, these results are slightly alarming. With cardboard, the virus lasts 24 hours. During the pandemic, most companies have added COVID-19 protection measures to their shipping process that reduces the risk of virus transmission. So if you were wondering, don't worry too much about bringing that delivery package inside, but make sure to wash your hands well afterwards! Most importantly (to this chapter as you will soon see) though, the new coronavirus lasted 4 hours on copper surfaces.

In another study, the authors found that the virus was not detected on wood and cloth fabric after 2 days. On glass and paper money, it lasted 4 days. Furthermore, they tested surgical masks and discovered the virus had survived on the exterior of the mask on the 7th day, which is notably longer than that of cloth fabric. What this highlights is the need to dispose of surgical masks correctly and discourage touching the face, regardless of whether we are wearing a surgical or cloth mask.

How Long SARS-2-CoV-2 Lasts on Surfaces (according to the NIH Study)	
Surface type	Maximum number of days virus survives on surface
Copper	4 hours
Cardboard	24 hours
Plastic	2-3 days
Stainless steel	2-3 days

Now you might be wondering what these studies have to do with door levers and handrails. Recall that Doramalen and colleagues discovered that the SARS-CoV-2 virus is deactivated within 4 hours on copper surfaces. When compared to the other hard surfaces tested, copper seems far superior. Wouldn't it then be beneficial to take advantage of these findings and use copper to produce more protective surfaces for the public?

How long SARS-CoV-2 lasts on surfaces (according to Chin et al., 2020)	
Surface type	Maximum number of days virus survives on surface
Paper (including tissue paper)	3 hours
Wood	2 days
Cloth	2 days
Glass	4 days
Paper money	4 days
Surgical masks	7 days

Researcher Bill Keevil has spent more than twenty years studying copper and its antimicrobial properties. His research shows copper can kill germs. Keevil has tested copper surfaces with the MERS and H1N1-linked viruses and observed that both were deactivated in minutes. After testing it on a coronavirus that causes the common cold, he found the same results. Keevil has since advocated for its wide use in the United Kingdom, and believes it is far undervalued and underused despite its effectiveness.

Keevil's research has far-reaching implications for high-risk areas. One study published in August discovered that viral DNA on a hospital bed rail can spread to 18 other surfaces (such as waiting room chairs) within 10 hours! Just imagine where the virus could land after a day! This could have detrimental outcomes, especially for patients at high risk for complications. But if copper is able to kill bacteria and viruses, it could be used to create doorknobs, bathroom fixtures, and bed rails that maximize patient and public safety.

The problem? Our sweat. It seems like a paradoxical problem, but sweat weakens copper's antibacterial properties. Sweat's main component is salt, making this the one instance in this book where salt is disadvantageous. When salt comes into contact with the metal, it prevents copper from deactivating bacteria. In a study by Bond and Lieu, a corrosive layer formed on copper surfaces within an hour of human contact. Now an hour may not sound terrible, but think about how many people touch sink handles, or handrails, and practice consistently good hand hygiene. If more people touch the fixture, the corrosive layer will become stronger and effectively make the use of copper pointless.

Another solution, without the drawbacks of copper, has been proposed during the pandemic. The company Outbreaker Solutions has been working with the University of Alberta to test a unique product for commonly touched surfaces: compressed sodium chloride (CSC). By subjecting typical salt particles to high pressure, salt transforms into a ceramic-like material capable of taking on any size and shape. So instead of a salt coating, entire surfaces can be made of salt.

You may be wondering though, what happens when the product gets wet? When salt is compressed, the process changes its innate properties. As a result, CSC does not instantly dissolve like table salt, and can be disinfected with typical (or hospital grade) cleaning products without breaking down. In terms of long-term efficiency, Outbreaker Solutions says their products should last 10-12 years.

Past research has shown that CSC has "strong antibacterial and antifungal effects, killing microbes in minutes". Compared to copper, CSC works quickly within seconds, rather than minutes. The science behind its effectiveness is similar to the salt coating developed by Choi and colleagues (described in chapter 12). When germs in the form of water droplets come into contact with CSC, the salt dissolves. As the salt dissolves, water evaporates and causes salt crystals to form. These salt crystals are pointy, and their sharp edges kill the virus.

In previous chapters, we explained how salt is capable of deactivating influenza strains. This project by Outbreaker Solutions hopes to find out whether salt can kill other viruses such as norovirus (a contagious virus that causes inflammation in the digestive tract) and SARS-CoV-2. If the project has successful outcomes, it means that we can implement CSC coating on hard surfaces (door levers, push plates, bed rails, handrails) in high-risk areas such as "hospitals, long-term care centres, cruise ships and the food processing industry, where outbreaks frequently occur". Since CSC is made from table salt, it is cheaper than the typical materials used for public fixtures, making it practical. The collaborative group hopes to secure more funding and get their solution implemented in public spaces.

While we can't change people and their behaviour, especially during a pandemic, we can improve our external conditions to better protect people in high-risk and high-traffic zones. In this chapter, we discussed the survival patterns of SARS-CoV-2 on various surfaces, and how researchers are experimenting with different materials to reduce them. Research is promising, but it should not diminish the importance of hand hygiene and surface disinfection. The COVID-19 virus can survive on our hands for several hours before degrading completely, and if you touch your face with a contaminated hand, you are making yourself vulnerable to infection. The safest way to prevent yourself from contracting the virus is to consistently wash your hands (for at least 20 seconds with soap) and regularly disinfect commonly touched surfaces.

Saline in other PPE

If someone had asked you in 2019 what the acronym PPE stood for, you probably would have struggled to answer their question. Now, in the age of the COVID-19 pandemic, practically everyone is aware of the term personal protective equipment (PPE) and frontline healthcare workers depend on said equipment to treat COVID-19 positive patients while minimizing their chance of catching the virus. However, you may be surprised to learn that the term also covers what many citizens and essential workers wear on a daily basis. PPE covers anything that is worn to provide a barrier between an individual and other objects or people with the purpose of the prevention of exposure to infectious diseases. This includes gloves, gowns, masks, respirators, goggles, face shields and masks with visors.

At the beginning of the pandemic, there became a world wide shortage of PPE because prior supply could not meet the sudden increase in demand. Furthermore, many individuals began hoarding items such as masks and gloves due to the panic of an impending shortage of these supplies, which contributed to the shortage of these items. Despite the World Health Organization shipping over 500 000 supplies to almost 50 countries, supplies continued to deplete at an extremely rapid pace. One item that is still urgently needed by some hospitals and clinics are N-95 masks, which have been recommended by the Center for Disease Control and Prevention (CDC) to not be worn by the public. This recommendation stems from the lack of N-95 masks available to medical professionals who require it the most. These masks require complex machinery to build and workers that are specialized in their production. The filter within the mask is made of thousands of thin fibres that are melted together. The material forming the fibres are not easy to obtain and their scarcity on top of the lack of manufacturers available to make the masks have created a worldwide shortage. An N-95 mask is meant to fit very snugly on one's face and is considered the most effective mask at filtering airborne particles. The shape of the mask forms a seal surrounding one's nose and mouth so no particles are able to enter. The downsides to this type of mask are that it can make it difficult to breathe and they were designed to be single use. However, limited reuse of the masks is currently being widely promoted in efforts to conserve supplies. The United States Food and Drug Administration has even issued Emergency Use Authorizations detailing the decontamination process that must be followed prior to their reuse.161 Surgical masks are also used by healthcare workers and are considered to be critical supplies, although amidst the COVID-19 pandemic many people are opting to use surgical masks to provide protection from the virus. A surgical mask tends to be more loose fitting and does not protect against very small particles but is effective at providing a barrier to block larger droplets that may contain viruses. These masks, along with cloth masks, are effective at preventing the general population from contracting COVID-19 as droplets are the main source of transmission of the virus. Unlike cloth masks, surgical masks are meant to be worn once and should be disposed of safely after use by placing it in a plastic bag and then the trash. At this time, the CDC recommends the general public to wear cloth masks which can be washed after use and can be made easily at home. These masks are to be worn anytime someone is unable to social distance and is in a public setting regardless of whether they have COVID-19 symptoms or not.

In order for researchers to create PPE that will successfully protect wearers against a given virus they must be aware of how it is transmitted and how contagious the virus is. The two major issues that are focused on when designing respirators or masks is fit and filtration. Both are extremely important as filtration refers to how many particles and what size of particles are able to penetrate through the mask. On the other hand, fit is necessary to ensure particles do not enter through areas where the mask does not fit snugly against the skin. Other PPE such as gowns, gloves, and eye protection are primarily designed to act as barriers between the skin and droplet spray or contact transmission.164

To test filtration efficiency, researchers will often use salt particles as a representation of viral droplets and particles. A study conducted by Oberg and Brosseau showed that the standard dental mask showed significantly higher penetration of salt particles (53-90%) than hospital masks (4-37%). Li et al. piloted a similar study in 2006 in which they used potassium chloride instead of sodium chloride as representation of the particles and the results showed 95% filtration efficiency. However, the former study by Oberg and Brosseau pointed out flaws within Li et al.'s research techniques as they did not follow the current accepted standard research method and were unable to accurately describe the techniques they used. Seeing as Li et al's methodology may have been flawed, it is difficult to ascertain the accuracy of their results.

Although PPE such as gloves and gowns are widely seen as an effective way for healthcare workers to protect themselves from contracting viruses, there is always a risk of transmission. Most of this risk arises when individuals are taking off PPE that may be contaminated. Due to the concern surrounding potential virus transmission while removing PPE during the Ebola outbreak in 2014-2016, the CDC recommended decontaminating PPE prior to removal. A study published in 2019 by Zegbeh Kpadeh-Rogers and others set out to determine whether this was truly a beneficial method at reducing the risk of virus transmission.166 They found that glove contamination prior to glove removal does in fact significantly reduce the risk of bacterial contamination. However, there was still bacteria detected when the decontaminated gloves were tested upon removal.166 This indicates that while decontaminating gloves can reduce the risk of health-care professionals accidentally contaminating themselves while removing them, it does not completely eliminate the risk. As for what decontaminant is most efficient, the study found that using wipes made with bleach (sodium hypochlorite) reduces the risk more than alcohol based wipes did. Despite these findings, it is necessary to note that while bleach is an effective disinfectant that can kill bacteria and viruses, it can also cause damage with prolonged use. Bleach can easily irritate the skin, airway, and mucous membranes and can be inactivated quickly in the presence of organic materials. Furthermore, it will often react quickly with other chemicals. Therefore, any recommended use of bleach should always inform individuals of all the necessary precautions that they must take if they elect to utilize the chemical.

For most people not in the medical field, how health-care workers decontaminate their gloves is less important than how one can use gloves effectively and disinfect surfaces in their homes amidst the COVID-19 pandemic. As of August 2020, public health officials are not recommending the use of gloves when running errands, simply because they can do more harm than good.167 If you are wearing gloves while running the errand, you are probably more likely to forget that you shouldn't be touching your face after touching other items, making it easier to spread germs. Furthermore, if you dispose of the gloves incorrectly and your bare hand touches the outside of the gloves you would be exposed to more germs than if you simply washed your hands with water and soap. However, if you do choose to wear gloves it's important to not touch the outside of the glove with your bare fingers when removing them and to not reuse gloves. Once you place the gloves in the trash it is still necessary to wash your hands with warm water and soap for at least 20 seconds. Although there is not much research regarding whether or not it is likely that COVID-19 will spread via groceries, many are taking the additional precaution of wiping down their groceries upon returning from the store. The best way to do this would be to wash all your boxes or cans of food with soap, throwing out all disposable packaging, and using disinfectant wipes to wipe down any other groceries. Following this step you should wipe down any surface that came into contact with the grocery bags or groceries themselves before washing your hands with warm water and soap. If you run out of all your disinfectant wipes, you can easily make a DIY disinfectant using some bleach and water. For every 5 mL of bleach that you add to your solution you should add 250 mL of water, thus creating a 0.1% solution of bleach. For most people, simply soap and water would suffice when cleaning surfaces in your household; however, if you are caring for someone in the home that has a confirmed case of COVID-19 it is necessary to use disinfectant.171 This is because disinfectant actually kills off the bacteria, while soap simply reduces the amount of bacteria. When possible, it is best to use a federally approved disinfectant, but the bleach solution will suffice in case of emergencies. Surfaces that are important to disinfect at least once a day if you are living with someone that has COVID-19 include all high-touch surfaces such as doorknobs, phones, bedside tables and remotes.171

While it seems the novel coronavirus has effectively made everyone an expert on personal protective equipment such as masks and gloves, we must continue to be cautious in our use of them. Paying attention to the possibility of contamination and correct disinfection procedures can greatly reduce risk of infection and ensure that our PPE is doing its job. No matter how carefully designed our masks, gloves, and face-shields are, they are only as effective as we allow them to be.

Antimicrobial Surface Development using Salt

The term salt, when mentioned in a day to day context usually refers to table salt, or what chemists know as NaCl, or sodium chloride. Sodium chloride itself is just positively charged sodium ions ionically bonded to negatively charged chloride ions. By extrapolating this information we come to the actual definition of salts used by scientists everywhere: a salt is a chemical compound formed of cations and anions held together by an ionic bond . Cations are positively charged ions, which have 1 or more electrons missing from its outer valence shell. Anions on the other hand are negatively charged ions that have 1 or more extra electrons in its outer valence shell. These two ions are held together by an ionic bond, which is weaker than covalent bonds but much stronger than hydrogen bonds`68. So with this new definition of salt, a whole new realm of possibilities opens up including the different salts that can be made and their own unique uses. Well, in terms of how many salts you can make, well there's quite literally hundreds, and different salts even fit into different categories168. For example, salts that produce hydroxide ions, OH-, when they're dissolved in water are known as alkali salts or you can have salts that are organic (contain carbon compounds) such as ethyl acetate168. So as you can see there's a multitude of salts out there, but a key one that is important in antimicrobial surface development are silver salts . Silver salts include silver nitrate, silver phosphates, silver chloride and more, and each has different properties and uses. But the main and key component to all of them is the silver, as silver is the key antimicrobial component in silver salt based technologies169.

Humans have known about silvers antibacterial properties for a long time now, in fact records date back the Greek and Roman empires. In fact, historically aristocrats were known to have blue skin and scientists today think it has to do with how much silver they used and kept around169. However, it was until the 1970s that scientists really began to understand how silver works and why it does. Today, we know there are 4 main methods of action of silver against microbes including bacteria and viruses (though the former more than the latter)169. However in both of these mechanisms silver must be in its ionic form, which is where silver salts come into play. Silver on its own is actually inert and does not possess antimicrobial properties. So silver salts such as silver nitrate, when added to water or electrically charged will cause the compound to split into two ionic molecules, one of them being ionic silver. Luckily, the insides (sometimes the outsides as well) of cells both human and bacterial contain water, thus when silver salts come into contact with it, silver ions are released. The first main mechanism through which silver acts is by bonding to thiol groups in enzymes and causing them to deactivate or change function. Thiol group are sulfur hydrogen side groups (-SH) found in certain amino acids and amides (such as cysteine an amino acid)169. In particular there are many membrane proteins that contain thiol groups, and silver ions can actually form stable bonds to the sulfur in the –SH group (forming S-Ag bonds). Silver tends to form these bonds with transmembrane proteins involved in ion transport and transmembrane energy generation, both of which are key processes . This disruption often leads to bacterial cell death, or at the minimum will restrict bacterial cell growth.

Silver can also enter cells and affect cellular respiration, by acting as a catalyst. A catalyst is a molecule or protein that speeds up a reaction, and in the case of silver it can catalyze the formation of disulfide bonds between two –SH groups. The disulfide bond formation will affect the secondary structure of the protein thereby affecting the proteins function, and thus potentially inactivating it . The second mode of action for silver is to affect the DNA of bacterial cells. The silver can intercalate between the two double stranded DNA strand, binding to the purine and pyrimidine bases, and disrupting the DNA itself2. Silver has also been seen having an effect on bacteria cell division, thereby inhibiting replication, thus stopping the growth of the bacterial colonies . Furthermore, silver ions also work by entering the cell and binding to ribosomes such that the messenger RNA (mRNA) cannot be transcribed, which means proteins cannot be created. The last mechanism through which silver works is by binding to the –SH group of cytochrome b, but not much is known about this and it is secondary to the methods previously mentioned172.

So, now that we know exactly how silver works it's important to understand what antimicrobial surfaces are and how silver salts relate to it. A big problem in hospitals and hospital-like settings is the spread of infection, in fact healthcare associated infections (HAIs) are a frequent challenge for patients and healthcare providers .

Masks and gloves can only do so much. Studies have found that in order to completely control bacterial and fungal growth in such settings, gurneys, stretchers and other equipment would need to be cleaned with harsh disinfectants every 2 hours. Moreover, bacteria that grow in hospital settings are more likely to become resistant to certain antibiotics, through natural evolution, due to constant use . As such it becomes a more and more prominent issue to find ways to keep equipment and materials clean and safe. This is where antimicrobial surfaces come into play . These surfaces are designed to kill bacterial cells, to clean layers of dead bacterial cells off and to keep surfaces clean, however managing to do all 3 at the same time with a high efficiency is extremely hard. That's where silver coated antimicrobial surfaces come into play175.

We already know how and why silver is used against the microbiome, now it's about the development of the surfaces. A clear example of antibacterial surface is SilverShield by Microban, a company that integrates silver phosphate into various materials such as polymers, ceramics and coatings . The coating is leach proof and doesn't wash off, but when in contact with water or a damp environment (a little goes a long way), silver ions will be released, thus killing the bacterial cells176. Another study done in 2017, showed that silver ion coatings (aka dissolved silver salts) can directly cause a decrease in bacterial growth, however the percentage of bacterial death is dependent on environment, application method and cleanness of the surface174. By looking at hospital bedside tables, and OR tables this study was able to show the direct application of this technology to a hospital based setting and how it can be beneficial174. Other products very similar to this are also being used in hospital and industrial settings, however all of these products can be expensive and inefficient at times. These issues become even more challenging when we talk about making surgical equipment antibacterial. Bacterial infections represent very serious complications in surgery, especially vascular surgery due to its increase in usage of artificial prostheses . A promising combination used to combat bacterial accumulation in vascular surgery is silver phosphate nanoparticles. Nanoparticles are particles that are between 1 and 100 nanometers in diameter, so very small. But because they're so small they can have unique properties. In terms of the silver phosphate nanoparticles, it's been shown that they have a high toxicity to bacteria and bacterial colonies, but a low toxicity to humans177. This makes it acceptable for use for surgical instruments thus helping to contain HAIs177.

Antimicrobial surfaces don't just refer to abiotic surfaces and objects, but can also be referring to wounds and burns on human skin. Burns, wounds and other abrasions on human skin can open the doors to various bacterial growth and infections . Having an open cavity allows the bacteria normally living on top of the skin access to the bloodstream and mucus layers through which it can enter the human body thus leading to infections and disease. For years now, standard dressing has not been enough to help combat most infections, which is why salt based dressings are now being used . Salts such as silver nitrate (0.5%) and silver sulphadiazine have been in use since the 1960s. However, a bacterial resistance to these substances was seen to take place in the late 90s, thus resulting in them being less common178. Today, as of 2011, Britain has authorized the use of silver nitrate (40-95%) for external use on warts, verruca, umbilical granulomas and cauterization178. The use of silver sulphadiazine (1%) (which is a combination of silver and sulphonamide) has also been approved to help combat infections in burn wounds, leg ulcers and pressure sores.

These secondary uses of silver provide additional surfaces that are likely to be contaminated with bacteria which we as humans can battle through ionic silver technology179. In addition to silver salts being so widely used in various different applications within the healthcare industry, silver has also recently been used to make silver nanoparticles. These silver nanoparticles are made from silver salts, usually silver nitrate . Silver nitrate is reduced using an aldehyde as a reducing agent and an amino acid as a catalyst, and through this reaction silver nanoparticles are formed usually between 70 and 350 nanometers wide.

This reaction of silver salts is extremely important because the next generation of antibacterial surfaces relies heavily on silver nanoparticles. Silver nanoparticles are more durable than ionic silver and are being more heavily researched and used150. A study in 2017 found that glass coated with silver nanoparticles could inhibit E.coli growth within one cycle and S.aureus growth within two cycles . These results clearly indicate the effect these particles have on bacterial inhibition and showcase the potential silver salts and silver nanoparticles have for future antimicrobial surfacing181.

Overall, although not the salt everyone knows so well, NaCl, silver salts are extremely useful bactericidal (bacteria killing) compounds. Through its 3 major mechanisms silver can cause bacterial death or inhibit bacterial growth, which are both important in containing infections2. Hospital based transmissions are very prevalent due to the fact that bacteria and microbes can last on surfaces for hours if not days depending on the class . The answer to this challenge is antimicrobial surfaces using a natural bactericidal such as ionic silver (aka silver salts)2. These can be used anywhere from hospital bed rails and tables to prosthetics for surgeries, or even burns and wound dressings183. Future research in this field is extremely important as with new technology and the creation of silver nanoparticles, our fight against the microbiome continues.

Conclusion

From wearing masks in public spaces to adhering to social distancing guidelines on a daily basis, the idea and feeling of normalcy in our society has drastically changed within the last few months. The initial news of the deadly spread of COVID-19 left citizens worldwide in a state of shock, confusion, and fear. With no guidelines or protocols in place to protect ourselves in the beginning stages of this developing virus, feelings of uncertainty surrounding our health began to waver as we were faced with difficult decisions on how to prepare ourselves for the unknown. As the WHO deemed COVID-19 a global health emergency in late January, and then a pandemic in March, people worldwide began to realize its severity, and that it should not be treated lightly as a common cold or the flu. As the weeks passed, the world spiraled into an economic decline as citizens worldwide faced the deadly consequences of the virus, including lost jobs, the closure of small businesses, evictions as a result of late rent payments, and cancelled internships for students and young adults entering the workforce. It is not an understatement to say that the last few months have been troubling, grief-stricken, and ever-changing, as life may never be the same as it was during pre-pandemic times.

COVID-19 has quite literally taken the world by a storm and the damage it's done worldwide, from a loss in global revenue to a rapid increase in the demand for PPE (personal protective equipment), poses a threat for our future. COVID-19 threatens to undo progress achieved in developing countries such as Rwanda and El Salvador, as they battle underdeveloped healthcare systems and losses in household incomes. Some of the longer-term risks of the coronavirus pose an even larger risk for countries with ongoing social justice and economic inequalities. The UN has said that "unless bold policy actions are taken by the international community" the least developed countries (LDC's) will face the wrath of this deadly disease to extremes we didn't think possible.

Here in Canada, we have relied on the government to minimize the health, economic, and social impacts of this rapidly evolving crisis. Funds such as the CERB (Canada Emergency Response Benefit) and the CESB (Canada Emergency Student Benefit) have helped millions of Canadians obtain a steady source of income, while also giving students and young adults opportunities to volunteer and work through federal job opportunities and the CSSG (Canadian Student Service Grant). Moreover, we have seen the direct impact of how uprooting one's daily activities and schedule can take a toll on an individual's mental health. Amidst grim government briefings and headlines profiling bleak outcomes, "experts have warned of an "echo pandemic" of mental illness coming in the wake of the COVID-19 outbreak".

As we begin to slowly recover from the economic strain and physical toll this pandemic has taken upon us, we have begun to realize the extent to which we rely on technology to communicate and educate ourselves on COVID-19 related news. However, we have also realized the extent to which news outlets and public figures embellish false claims and amplify certain misconceptions about the virus. These misconceptions spread like wildfire; some would even say as quickly as the virus itself. Some of the most common misconceptions regarding COVID-19 are that ingesting hydroxychloroquine (the main, active ingredient in bleach) or other disinfectants have clinical benefits in treating COVID-19 and have the ability to 'cleanse' the body. Various accredited institutions including John Hopkins University and companies such as Reckitt Benckiser—the parent company of Lysol— have strongly condemned and fact-checked such falsified practices and methods. It is vital that the public allows researchers to do their job and guide us on ways to prevent the spread rather than us becoming self-proclaimed professionals. With that being said, using credible sources to stay updated on the developments of this global health crisis is vital to ensure our safety and health. This book has been one of those many ways that complies relevant and reliable research for readers to become educated on salt and its role in this pandemic.

Other DIY solutions that have been circling the Internet revolve around the use of salt. Debunking various salt related myths that have been plastered all over blog sites, Facebook group chats, and news outlets was a crucial part of this book to establish that while salt has antibacterial properties, it is not beneficial to consume an excess of salt to cleanse the body or prevent bacteria from entering the body. This can lead to a vast amount of medical-related problems including hypertension (high blood pressure). Similarly, it is detrimental to entirely eliminate salt from the diet as well as it can lead to a variety of health issues including hyponatremia, the condition defined as an extreme loss of sodium that can subsequently trigger other symptoms ranging from muscle cramps to nausea. From a health and lifestyle point of view, salt serves as a vital source of electrolytes with chloride ions regulating both blood pH and blood pressure. Due to perspiration or excretion via urine, salt must be replenished through one's diet to remain healthy and to have all internal systems in balance.

We hope by now, as readers, you have gained a thorough understanding of salt; not only from a nutrition perspective, but also from a healthcare and research point of view. Salt is an underappreciated element, both in the kitchen and in research. Saline, a mixture of salt and water, has different concentrations of salt within the solution. As discussed, it is one of the most widely used and recognized tools in healthcare settings. From wound and nasal irrigations to prevent infection or to saline IV's (intravenous lines) that replenish lost fluids and deliver medications, salt is an essential component when treating patients and clients. Saline has also played a large role in the context of COVID-19 with ongoing trials focused on saline-infused masks to decrease transmission and act as another barrier to prevent other bacteria from entering our bodies. On the other hand, salt and its microbial properties allow for it to be a key disinfectant in cleaning supplies and killing bacteria or microorganisms. While some of its mechanisms are still being understood by scientists, its role in research studies is paving the way forward to discovering new treatment plans for those diagnosed, as well as its role in preventative measures to decrease the transmission and spread of the virus between people. It is vital to increase awareness of the research and trials being done to develop current prototypes of PPE (personal protective equipment) such as salt-infused masks and preventative measures to decrease transmission, such as the compressed sodium chloride (CSC) installed on high-touch surfaces. Salt is also being used as an aerosol in halotherapy, an upcoming non-medicative intervention designed to alleviate symptoms of various respiratory diseases and ailments.

It is important to acknowledge how far research has come within the last few years and how salt has played a role in the SARS outbreak, the H1N1 virus outbreak, and now, COVID-19. Each new approach, study, trial, and prototype brings us a step closer to finding a treatment and cure for the global health crisis of the century. Salt and its extensive history, dating back to its discovery far earlier than any historical record, has proven its worth as an essential component in our world of food, disinfectants, and preservatives. Whole cities, towns, and civilizations were built based on the availability of salt, as it was considered to be a vital aspect of human life given its widely accessible and inexpensive nature. While this may seem absurd, it proves to us that salt and its multifaceted roles over decades and centuries have emerged to be effective.

When we began compiling this book as writers, we were amazed at the complexity of a simple kitchen ingredient and what it could be used towards. We learnt more than we expected through weeks of research and writing, and it is our hope that we are able to spread this knowledge and our findings to people worldwide. We hope you thoroughly enjoyed reading through our book and learned a thing or two along the way about salt!

References

Action on Salt: Kidney Disease and Kidney Stones. (n.d.). Retrieved from http://www.actiononsalt.org.uk/salthealth/factsheets/kidney/

Armstrong, M., & Richter, F. (2019, August 06). Infographic: China Isn't Alone In Its Salt Consumption Problem. Retrieved July 29, 2020, from https://www.statista.com/chart/18937/salt-consumption/

A salt-coated mask that kills viruses? Alberta researchers working on it | CTV News. Accessed August 15, 2020. https://www.ctvnews.ca/health/a-salt-coated-mask-that-kills-viruses-alberta-researchers-working-on-it-1.4798138

Badrick, A. (2018). Control of Blood Pressure - Short and Long Term - TeachMePhysiology. Retrieved 21 July 2020, from https://teachmephysiology.com/cardiovascular-system/circulation/control-blood-pressure/

Bar-Yoseph, R., Kugelman, N., Livnat, G., Gur, M., Hakim, F., Nir, V., & Bentur, L. (2016). Halotherapy as asthma treatment in children: a randomized, controlled, prospective pilot study. Pediatric Pulmonology, 52(5), 580-587. https://doi.org/10.1002/ppul.23621

Ben Guarino, C. J. (2020, June 13). Spate of new research supports wearing masks to control coronavirus spread. Retrieved August 15, 2020, from https://www.washingtonpost.com/health/2020/06/13/spate-new-research-supports-wearing-masks-control-coronavirus-spread/

Berg, S. (2020). Do your patients believe these 7 myths about salt? (2019) Retrieved 21 July 2020, from https://www.ama-assn.org/delivering-care/hypertension/do-your-patients-believe-these-7-myths-about-salt

Blood Pressure : Salt's effects on your body. (2020). Retrieved 21 July 2020, from http://www.bloodpressureuk.org/microsites/salt/Home/Whysaltisbad/Saltseffects#:~:text=Eating%20salt%20raises%20the%20amount,vessels%20leading%20to%20the%20kidneys

Blood Pressure: Salt's Effects on Your Body. (n.d.). Retrieved from http://www.bloodpressureuk.org/microsites/salt/Home/Whysaltisbad/Saltseffects

Bogart, N. (2020, April 03). Setting routines, picking up hobbies key to managing mental health during COVID-19, experts say. Retrieved from https://www.ctvnews.ca/health/coronavirus/setting-routines-picking-up-hobbies-key-to-managing-mental-health-during-covid-19-experts-say-1.4881576

Bogart, N. (2020, February 05). A salt-coated mask that kills viruses? Alberta researchers working on it. Retrieved from https://www.ctvnews.ca/health/a-salt-coated-mask-that-kills-viruses-alberta-researchers-working-on-it-1.4798138?fbclid=IwAR0A99QlPOieEJPYsnBnZ1KmKkfqvg_a4wdIILWa20XCQkfk6VR3XjGfAlw

Bond, J. W., & Lieu, E. (2014). Electrochemical behaviour of brass in chloride solution concentrations found in eccrine fingerprint sweat. Applied Surface Science, 313, 455-461. https://doi.org/10.1016/j.apsusc.2014.06.005

Brabraw, K. (2019, Jan 3). This is how long cold and flu germs can live on surfaces like doorknobs and subway poles. Health. Retrieved from https://www.health.com/condition/cold-flu-sinus/flu-virus-live-on-surfaces

Brown, K. V., & Sink, J. (2020, April 23). Trump's Idea to Inject Disinfectant Alarms Medical Experts. Retrieved from https://www.bloomberg.com/news/articles/2020-04-23/coronavirus-dies-fastest-under-light-warm-and-humid-conditions

Canada, H. (2013). Sodium: the basics - Canada.ca. Retrieved 21 July 2020, from https://www.canada.ca/en/health-canada/services/nutrients/sodium/sodium-basics.html?wbdisable=true

Canada, H. (2020, July 21). Government of Canada. Retrieved August 1, 2020, from https://www.canada.ca/en/health-canada/services/drugs-health-products/covid19-industry/medical-devices/personal-protective-equipment/overview.html

Canada, P. (2020, July 03). Government of Canada. Retrieved August 16, 2020, from https://www.canada.ca/en/public-health/services/publications/diseases-conditions/how-to-care-for-person-with-covid-19-at-home-advice-for-caregivers.html1.

Canada E and CC. Water: frequently asked questions. aem. Published April 1, 2009. Accessed August 15, 2020. https://www.canada.ca/en/environment-climate-change/services/water-overview/frequently-asked-questions.html

Center for Devices and Radiological Health. (n.d.). N95 Respirators, Surgical Masks, and Face Masks. Retrieved August 16, 2020, from https://www.fda.gov/medical-devices/personal-protective-equipment-infection-control/n95-respirators-surgical-masks-and-face-masks

Center of Disease and Preventions. (2020, July 15). Frequently Asked Questions. Retrieved from https://www.cdc.gov/coronavirus/2019-ncov/faq.html

Centers for Disease Control and Prevention. (2009, December 5). Treat other flu symptoms. Retrieved from https://www.cdc.gov/h1n1flu/homecare/treatsymptoms.htm

Centers for Disease Control and Prevention. (2010, February 10). 2009 H1N1 flu ("Swine Flu") and you. Retrieved from https://www.cdc.gov/h1n1flu/qa.htm

Centers for Disease Control and Prevention. (2019, May 8). 2009 H1N1 timeline. Retrieved from https://www.cdc.gov/flu/pandemic-resources/2009-pandemic-timeline.html

Centers for Disease Control and Prevention. (2020, July 27). What is the difference between Influenza (Flu) and COVID-19? Retrieved from https://www.cdc.gov/flu/symptoms/flu-vs-covid19.htm

Chin, A. W. H., Chu, J. T. S., Perera, M. R. A., Hui, K. P. Y., Yen, H., Chan, M. C. W., Peiris, M., & Poon, L. L. M. (2020). Stability of SARS-CoV-2 in different environmental conditions. The Lancet, 1(1), E10. https://doi.org/10.1016/S2666-5247(20)30003-3

Commissioner, O. (n.d.). Is Rinsing Your Sinuses With Neti Pots Safe? Retrieved August 16, 2020, from https://www.fda.gov/consumers/consumer-updates/rinsing-your-sinuses-neti-pots-safe

COVID-19 and the least developed countries | Department of Economic and Social Affairs. (2020, May 1). Retrieved from https://www.un.org/development/desa/dpad/publication/un-desa-policy-brief-66-covid-19-and-the-least-developed-countries/

COVID-19 Mythbusters. (n.d.). Retrieved August 1, 2020, from https://www.who.int/emergencies/diseases/novel-coronavirus-2019/advice-for-public/myth-busters

COVID-19 Mythbusters. (n.d.). Retrieved from https://www.who.int/emergencies/diseases/novel-coronavirus-2019/advice-for-public/myth-busters

Cunningham, A. (2020, April 26). COVID-19 killer? Local doctors develop salt filters for masks to fight coronavirus. Retrieved from http://jamaica-gleaner.com/article/lead-stories/20200426/covid-19-killer-local-doctors-develop-salt-filters-masks-fight#slideshow-1

Does Gargling with Saltwater Really Help a Sore Throat? (2019, October 15). Retrieved June 29, 2020, from https://albanyentandallergy.com/does-gargling-with-saltwater-really-help-a-sore-throat/

Doremalen, N., Morris, D. H., Holbrook, M. G., Gamble, A., Williamson, B. N., Tamin, A., Harcourt, J. L., Thornburg, N. J., Gerber, S. I., Lloyd-Smith, J. O., Wit, E., & Munster, V. J. (2020). Aerosol and surface stability of SARS-CoV-2 as compared with SARS-CoV-1. New England Journal of Medicine, 382(16), 1564-1567. https://doi.org/10.1056/NEJMc2004973

electrolysis | Definition, Uses, & Facts | Britannica. Accessed August 15, 2020.

Elkins, M. R., Robinson, M., Rose, B. R., Harbour, C., Moriarty, C. P., Marks, G. B., Belousova, E. G., Xuan, W., & Bye, P. T. P. (2006). A controlled trial of long-term inhaled hypertonic saline in patients with cystic fibrosis. New England Journal of Medicine, 354(3), 229-240. https://doi.org/10.1056/NEJMoa043900

ELVIS COVID-19: About the Study. (2020, June 24). Retrieved from https://www.ed.ac.uk/usher/elvis-covid-19/about-the-study

Feyrer, J., Politi, D., & Weil, D. (2013). The Cognitive Effects of Micronutrient Deficiency: Evidence from Salt Iodization in the United States. doi: 10.3386/w19233

Get the Scoop on Sodium and Salt. (n.d.). Retrieved from https://www.heart.org/en/healthy-living/healthy-eating/eat-smart/sodium/sodium-and-salt

Goodwin, N. (2020, March 4). Some supplies running low as coronavirus fears grow. Canadian Broadcasting Corporation. Retrieved from https://www.cbc.ca/news/canada/ottawa/ottawa-coronavirus-supplies-1.5484074

Graphene-based sieve turns seawater into drinking water - BBC News. Accessed August 15, 2020. https://www.bbc.com/news/science-environment-39482342

Gunell M, Haapanen J, Brobbey KJ, et al. Antimicrobial characterization of silver nanoparticle-coated surfaces by "touch test" method. Nanotechnol Sci Appl. 2017;10:137-145. doi:10.2147/NSA.S139505

Haris, N. (2018, November 27). How to Reduce Sodium in Chinese Food. Retrieved July 29, 2020, from https://healthyeating.sfgate.com/reduce-sodium-chinese-food-5965.html

He, X., Lau, E. H. Y., Wu, P., Deng, X., Wang, J., Hao, X., Lau, Y. C., Wong, J. Y., Guan, Y., Tan, X., Mo, X., Chen, Y., Liao, B., Chen, W., Hu, F., Zhang, Q., Zhong, M., Wu, Y., Zhao, L., Zhang, F., Cowling, B. J., Li, F., & Leung, G. M. (2020). Temporal dynamics in viral shedding and transmissibility of COVID-19. Nature Medicine, 26, 672-675. https://doi.org/10.1038/s41591-020-0869-5

Health Canada releases a report on sodium consumption levels in Canada. (2018, July 23). Retrieved from https://www.newswire.ca/news-releases/health-canada-releases-report-on-sodium-consumption-levels-in-canada-688912541.html

Healthline, S. R. (2020, April 07). Here's How to Clean Your Groceries During the COVID-19 Outbreak. Retrieved August 15, 2020, from https://www.ecowatch.com/groceries-coronavirus-health-2645656475.html?rebelltitem=4#rebelltitem4

Hedman, J., Hugg, T., Sandell, J., & Haahtela, T. (2006). The effect of salt chamber treatment on bronchial hyperresponsiveness in asthmatics [Abstract]. European journal of allergy and clinical immunology, 61(5), 605-610. https://doi.org/10.1111/j.1398-9995.2006.01073.x

Heisler, Y. (2020, July 7). How long will we have to wear face masks in public? The answer isn't good. Boy Genius Report. Retrieved from https://bgr.com/2020/07/07/face-masks-coronavirus-how-long-wear-them-2021/

Helmenstine, A. M. (2019). What Exactly Is in Table Salt? Retrieved from https://www.thoughtco.com/what-is-table-salt-604008

Henney, J. E., Taylor, C. L., & Boon, C. S. (Eds.). (2010, January). Institute of Medicine (US) Committee on Strategies to Reduce Sodium Intake in the United States. Retrieved from https://www.ncbi.nlm.nih.gov/books/NBK50952/

High-salt diet affects the brains of mice. (2018, February 13). Retrieved from https://www.nih.gov/news-events/nih-research-matters/high-salt-diet-affects-brains-mice

Higuera, V. (2016, October 11). Disease Transmission: Direct Contact vs. Indirect Contact. Retrieved August 15, 2020, from https://www.healthline.com/health/disease-transmission

Homemade saline solution that could help abate COVID-19. (2020, March 30). Retrieved from https://baledoneen.com/blog/homemade-saline-solution-that-could-help-abate-covid-19/

How does our sense of taste work? (2016, August 17). Retrieved from https://www.ncbi.nlm.nih.gov/books/NBK279408/

How To Turn Salt Water To Drinking Water - Apex. Accessed August 15, 2020. https://apexwaterfilters.com/how-to-turn-salt-water-to-drinking-water/

How to use baths to manage your eczema. (n.d.). Retrieved from https://nationaleczema.org/eczema/treatment/bathing/

Human Coronavirus Types. (2020, February 15). Retrieved from https://www.cdc.gov/coronavirus/types.html

Institute of Medicine (US) Committee on Personal Protective Equipment for Healthcare Personnel to Prevent Transmission of Pandemic Influenza and Other Viral Respiratory Infections: Current Research Issues. (1970, January 01). Designing and Engineering Effective PPE. Retrieved August 16, 2020, from https://www.ncbi.nlm.nih.gov/books/NBK209586/

Intake I of M (US) C on S to RS, Henney JE, Taylor CL, Boon CS. Preservation and Physical Property Roles of Sodium in Foods. National Academies Press (US); 2010. Accessed August 15, 2020. https://www.ncbi.nlm.nih.gov/books/NBK50952/

Jon CohenMar. 27, 2., Lucy HicksAug. 12, 2., Lucy HicksJul. 31, 2., Charlotte

HartleyJul. 30, 2., Charlotte HartleyJul. 29, 2., & Charlotte HartleyJul. 24, 2. (2020, March 27). Not wearing masks to protect against coronavirus is a 'big mistake,' top Chinese scientist says. Retrieved August 1, 2020, from https://www.sciencemag.org/news/2020/03/not-wearing-masks-protect-against-coronavirus-big-mistake-top-chinese-scientist-says

K. M. (2020, April 20). Coronavirus: Should You Be Wearing Gloves When You Leave Home? Retrieved August 15, 2020, from https://www.houstonmethodist.org/blog/articles/2020/apr/coronavirus-should-you-be-wearing-gloves-when-you-leave-home/

Kennedy, K. (2017). 28 Canned Soups, Broths, and Stocks Under 500mg of Sodium | Everyday Health. Retrieved 21 July 2020, from https://www.everydayhealth.com/diet-and-nutrition/canned-soups-broths-stocks-under-500mg-sodium/

Kim, A., Andrew, S., & Froio, J. (2020, August 12). These are the states requiring people to wear masks when out in public. Retrieved August 1, 2020, from https://www.cnn.com/2020/06/19/us/states-face-mask-coronavirus-trnd/index.html

Knowledge and Practices Regarding Safe Household Cleaning and Disinfection for COVID-19 Prevention - United States, May 2020. (2020, June 11). Retrieved August 16, 2020, from https://www.cdc.gov/mmwr/volumes/69/wr/mm6923e2.htm?s_cid=mm6923e2_w

Kpadeh-Rogers, Z., Robinson, G., Alserehi, H., Morgan, D., Harris, A., Herrera, N., Leekha, S. (2019, September 13). Effect of Glove Decontamination on Bacterial Contamination of Healthcare Personnel Hands. Retrieved August 16, 2020, from https://www.ncbi.nlm.nih.gov/pmc/articles/PMC6761364/

Kurlansky M. Salt : A World History. 1st ed. Penguin Books; 2003.

Leffler, C., Ing, E., Lykins, J., Hogan, M., McKeown, C., & Grzybowski, A. (2020, January 01). Association of country-wide coronavirus mortality with demographics, testing, lockdowns, and public wearing of masks. Update August 4, 2020. Retrieved August 1, 2020, from https://www.medrxiv.org/content/10.1101/2020.05.22.20109231v5

Lehman, S. (2019, July 15). Can Consuming Too Little Sodium Cause Problems? Retrieved from https://www.verywellfit.com/what-happens-if-i-dont-consume-enough-sodium-2507757

Leonard, J. (2018, July 30). Does pink Himalayan salt have any health benefits? Medical News Today. Retrieved from https://www.medicalnewstoday.com/articles/315081

Liem, D. G., Miremadi, F., & Keast, R. S. (2011, June). Reducing sodium in foods: The effect on flavor. Retrieved from https://www.ncbi.nlm.nih.gov/pmc/articles/PMC3257639/

Lotz, S. (2015, August 28). Benefits of salt therapy. Organic spa magazine. Retrieved from https://www.organicspamagazine.com/benefits-of-salt-therapy/

Low, D. E., & McGeer, A. (2010). Pandemic (H1N1) 2009: assessing the response. Canadian Medical Association journal, 182(17), 1874–1878. Retrieved from https://www.ncbi.nlm.nih.gov/pmc/articles/PMC2988536/

Low Sodium Diet & Low Sodium Foods. (2019). Retrieved 21 July 2020, from https://my.clevelandclinic.org/health/articles/15426-sodium-controlled-diet

Maillard J-Y, Hartemann P. Silver as an antimicrobial: Facts and gaps in knowledge. Crit Rev Microbiol. 2012;39. doi:10.3109/104084 1X.2012.713323

McDermott, A. (2020). Halotherapy: Uses, Benefits, and Risks. Retrieved 21 July 2020, from https://www.healthline.com/health/halotherapy#methods

Mcneil, D. G. (2009, May 6). Debating the wisdom of 'Swine flu parties'. The New York Times. Retrieved from https://www.nytimes.com/2009/05/07/world/americas/07party.html

Mena, I., Nelson, M. I., Quezada-Monroy, F., Dutta, J., Cortes-Fernández, R., Lara-Puente, J. H., Castro-Peralta, F., Cunha, L. F., Trovão, N. S., Lozano-Dubernard, B., Rambaut, A., Bakal, H., García-Sastre, A. (2016). Origins of the 2009 H1N1 influenza pandemic in swine in Mexico. eLife, 5, e16777. https://doi.org/10.7554/eLife.16777

Morrison, J. (2020, April 14). Copper's virus-killing powers were known even to the ancients. Smithsonian magazine. Retrieved from https://www.smithsonianmag.com/science-nature/copper-virus-kill-180974655/

N95 Masks vs. Surgical Masks vs. Cloth Masks - Grainger KnowHow. (n.d.). Retrieved August 16, 2020, from https://www.grainger.com/know-how/

health/airborne-contaminants noise-hazards/respiratory-protection/kh-n95-masks-vs-surgical-masks-vs-cloth-masks

Najiya, H., & Cole, J. (n.d.). Salt. University of Waterloo. Retrieved from https://uwaterloo.ca/earth-sciences-museum/resources/detailed-rocks-and-minerals-articles/salt

Oberg, T., & Brosseau, L. M. (2008). Surgical mask filter and fit performance. American Journal of Infection Control, 36(4), 276-282. doi:10.1016/j.ajic.2007.07.008

Outbreaker Solutions. (n.d.). Efficacy Testing. Retrieved from https://www.outbreaker.ca/about

Outbreaker Solutions. (n.d.). Retrieved from https://www.outbreaker.ca

Outbreaker Solutions - CSC Antimicrobial. (2020, April 21). Outbreaker Solutions CSC Antimicrobial: FAQs [Video]. Youtube. Retrieved from https://www.youtube.com/watch?v=ljrbhNPWuLU&t=17s

Pappas, S. (2020, June 02). Do face masks really reduce coronavirus spread? Retrieved August 15, 2020, from https://www.livescience.com/are-face-masks-effective-reducing-coronavirus-spread.html

Park, C. (2020, March 16). Coronavirus: Saltwater spray infects 46 church-goers in South Korea. South China Morning Post. Retrieved from https://www.scmp.com/week-asia/health-environment/article/3075421/coronavirus-salt-water-spray-infects-46-church-goers

Petit, G., Jury, V., Lamballerie, M., Duranton, F., Pottier, L., & Martin, J. (2019). Salt Intake from Processed Meat Products: Benefits, Risks and Evolving Practices. Comprehensive Reviews In Food Science And Food Safety, 18(5), 1453-1473. doi: 10.1111/1541-4337.12478

Post, A., Dullaart, R. P., & Bakker, S. J. (2020). Is low sodium intake a risk factor for severe and fatal COVID-19 infection? European Journal of Internal Medicine, 75, 109. doi:10.1016/j.ejim.2020.04.003

Publishing, H. (2012). Ask the doctor: Exercise and sodium - Harvard Health. Retrieved 21 July 2020, from https://www.health.harvard.edu/staying-healthy/exercise-and-sodium

Quan, F., Rubino, I., Lee, S., Koch, Brendan., & Choi, H. (2017). Universal and reusable virus deactivation system for respiratory protection. Scientific Reports, 7(39956). https://doi.org/10.1038/srep39956

Question Corner: Saltwater gargling. (2016, July 12). Retrieved August 16, 2020, from https://www.thehindu.com/sci-tech/science/question-corner-saltwater-gargling/article4051452.ece

Ramalingam, S., Graham, C., Dov, J., Morrice, L., & Sheikh, A. (2019). A pilot, open labelled, randomised controlled trial of hypertonic saline nasal irrigation and gargling for the common cold. Scientific Reports, 9(1), 1015. https://doi.org/10.1038/s41598-018-37703-3

Ramalingam, S., Graham, C., Dove, J., Morrice, L., & Sheikh, A. (2020). Hypertonic saline nasal irrigation and gargling should be considered as a treatment option for COVID-19. Journal of Global Health, 10(1). doi:10.7189/jogh.10.010332

Rawlinson, S., Ciric, L., & Cloutman-Green, E. (2020). COVID-19 pandemic – let's not forget surfaces. The Journal of Hospital Infection, 105(4), 790-791. https://doi.org/10.1016/j.jhin.2020.05.022

Reckitt Benckiser. (2020, April 24). Improper use of disinfectants. Retrieved from https://www.rb.com/media/news/2020/april/improper-use-of-disinfectants/

Robbins, R. (2013, March 19). Epithelial Cells in Urine: Types, Test Results, Causes, and More. Retrieved July 29, 2020, from https://www.healthline.com/health/epithelial-cells-in-urine

Roman Times | Salt History. The Salt Association. Accessed July 19, 2020. https://www.saltassociation.co.uk/education/salt-history/roman-times/

Saba, R. (2020, February 11). The secret ingredient in this face mask that could prevent the next coronavirus? A dash of salt. The Star. Retrieved from https://www.thestar.com/news/canada/2020/02/11/salt-is-the-secret-ingredient-in-these-face-masks-that-could-prevent-spread-of-next-coronavirus.html

salt | Definition & Properties. Encyclopedia Britannica. Accessed August 15, 2020. https://www.britannica.com/science/salt-acid-base-reactions

Salt and Sodium. (2019, December 16). Retrieved from https://www.hsph.harvard.edu/nutritionsource/salt-and-sodium/

Salt in Ancient Times. Accessed July 19, 2020. http://users.rowan.edu/~mcinneshin/101/wk07/salt.htm

Salt reduction. (2020, April 29). Retrieved from https://www.who.int/news-room/fact-sheets/detail/salt-reduction

Salt Therapy Association. (2018). Types of salt therapy. Retrieved from https://www.salttherapyassociation.org/types-of-salt-therapy

Sandilands, T. (n.d.). How a Salt Water Mouth Rinse Benefits Oral Health. Retrieved from https://www.colgate.com/en-us/oral-health/conditions/mouth-sores-and-infections/how-salt-water-mouth-rinse-benefits-oral-health-1214

SaltWorks. (n.d.). History of salt. Retrieved from https://seasalt.com/history-of-salt

ScienceDaily. (2017, January 5). New surgical mask doesn't just trap viruses, it renders them harmless. Retrieved from https://www.sciencedaily.com/releases/2017/01/170105160228.htm

Sea Water vs. Saline: Why Not All Salty Water Is Created Equal. (2019, July 02). Retrieved from https://ispyphysiology.com/2019/07/10/sea-water-vs-saline-why-not-all-salty-water-is-created-equal/

Secon, H. (2020, February 08). A biomedical engineer created a mask coated in salt that he says could neutralize viruses like the coronavirus in 5 minutes. Retrieved August 16, 2020, from https://www.businessinsider.com/mask-coated-in-salt-neutralizes-viruses-like-coronavirus-2020-2
Shaking up the Salt Myth: The Human Need for Salt. Chris Kresser. Published April 13, 2012. Accessed July 19, 2020. https://chriskresser.com/shaking-up-the-salt-myth-the-human-need-for-salt

Shaking up the Salt Myth: The Human Need for Salt. Chris Kresser. Published April 13, 2012. Accessed July 19, 2020. https://chriskresser.com/shaking-up-the-salt-myth-the-human-need-for-salt/

Shmerling, R. H. (2018, September 28). Drip bar: Should you get an IV on demand? Harvard Health Publishing. Retrieved from https://www.health.harvard.edu/blog/drip-bar-should-you-get-an-iv-on-demand-2018092814899

Shortage of personal protective equipment endangering health workers worldwide. (n.d.). Retrieved August 1, 2020, from https://www.who.int/news-room/detail/03-03-2020-shortage-of-personal-protective-equipment-endangering-health-workers-worldwide

Silver as an Antimicrobial Agent - microbewiki. Accessed August 15, 2020. https://microbewiki.kenyon.edu/index.php/Silver_as_an_Antimicrobial_Agent#Mechanism_of_action

Slotkin, J. (2020, April 25). NYC Poison Control sees uptick in calls after Trump's disinfectant comments. National Public Ratio. Retrieved from https://www.rb.com/media/news/2020/april/improper-use-of-disinfectants/

Sodium Intake of Canadians in 2017. (2018, August 14). Retrieved from https://www.canada.ca/en/health-canada/services/publications/food-nutrition/sodium-intake-canadians-2017.html

Sparks, M. A., Crowley, S. D., Gurley, S. B., Mirotsou, M., & Coffman, T. M. (2014). Classical Renin-Angiotensin system in kidney physiology. Comprehensive Physiology, 4(3), 1201–1228. https://doi.org/10.1002/cphy.c130040

Stax, O. (2013). 10.3 Muscle Fiber Contraction and Relaxation. Retrieved 21 July 2020, from https://opentextbc.ca/anatomyandphysiology/chapter/10-3-muscle-fiber-contraction-and-relaxation/

Streit, L. (2019). The 9 Healthiest Types of Cheese. Retrieved 21 July 2020, from https://www.healthline.com/nutrition/healthiest-cheese#section1

Tai, Z., & Sun, T. (2011, May 18). The rumouring of SARS during the 2003 epidemic in China. Retrieved August 16, 2020, from https://onlinelibrary.wiley.com/doi/full/10.1111/j.1467-9566.2011.01329.x

Tan, M. (2020, June 24). Salt: China's deadly food habit. Retrieved July 29, 2020, from https://theconversation.com/salt-chinas-deadly-food-habit-120201

Tanderson. How Does a Swimming Pool Salt Chlorine Generator Work? AutoPilot Website. Published January 17, 2014. Accessed August 15, 2020. https://autopilot.com/how-does-a-swimming-pool-salt-chlorine-generator-work/

Taylor, R. S., Ashton, K. E., Moxham, T., Hooper, L., & Ebrahim, S. (2011). Reduced dietary salt for the prevention of cardiovascular disease: a meta-analysis of randomized controlled trials (Cochrane review). American journal of hypertension, 24(8), 843–853. https://doi.org/10.1038/ajh.2011.115

The American Lung Association. (2016, June 9). Promising or placebo? Halo salt therapy: Resurgence of a salt cave spa treatment. Retrieved from https://www.lung.org/blog/promising-placebo-salt-halotherapy

The History and Benefits of Saline Water. (2015, November 25). Retrieved from https://advancedtissue.com/2015/11/the-benefits-of-saline-water/

The Salt Suite. (n.d.). Retrieved from https://www.thesaltsuite.com/respiratory

The Science Behind Salt. Morton Salt. Accessed August 15, 2020. https://www.mortonsalt.com/article/how-salt-works-2/

The science of chlorine-based disinfectant. (2013, December 11). Retrieved from https://www.cleanroomtechnology.com/news/article_page/The_science_of_chlorine-based_disinfectant/93824

Thompson A. New Graphene Filter Could Remove Salt From Saltwater. Popular Mechanics. Published September 5, 2017. Accessed August 15, 2020. https://www.popularmechanics.com/science/environment/news/a28065/graphene-filter-salt-from-saltwater/

Timm, J. C. (2020, April 23). 'It's irresponsible and it's dangerous': Experts rip Trump's idea of injecting disinfectant to treat COVID-19. Retrieved from https://www.nbcnews.com/politics/2020-election/it-s-irresponsible-it-s-dangerous-experts-rip-trump-s-n1191246?cid=sm_npd_nn_tw_ma

Tonog, P., & Lakhkar, A. D. (2020). Normal Saline [eBook edition]. StatPearls Publishing. Retrieved from https://www.ncbi.nlm.nih.gov/books/NBK545210/

Torborg, L. (2016). Mayo Clinic Q and A: Sea salt and sufficient iodine intake. Retrieved 21 July 2020, from https://newsnetwork.mayoclinic.org/discussion/mayo-clinic-q-and-a-sea-salt-and-sufficient-iodine-intake/#:~:text=Unfortified%20sea%20salt%20contains%20only,iodine%20per%20gram%20of%20salt.

U.S. Food and Drug Administration. (2018, March 23). Influenza A (H1N1)

2009 Monovalent Vaccines Descriptions and Ingredients. Retrieved from https://www.fda.gov/vaccines-blood-biologics/vaccines/influenza-h1n1-2009-monovalent-vaccines-descriptions-and-ingredients

Use of disinfectants: Alcohol and bleach. (1970, January 01). Retrieved August 16, 2020, from https://www.ncbi.nlm.nih.gov/books/NBK214356/

Vega, L. (2020, April 21). Door levers and handrails made of compressed salt could be effective against viruses such as COVID-19. Folio. Retrieved from

Vega, L. (2020, April 24). Door levers and handrails made of compressed salt could be effective against viruses such as COVID-19. Retrieved from https://www.folio.ca/door-levers-and-handrails-made-of-compressed-salt-could-be-effective-against-viruses-such-as-covid-19/

What Is Hypochlorous Acid. (2020, July 09). Retrieved from https://www.forceofnatureclean.com/what-is-hypochlorous-acid/

Why some Americans won't wear face masks, in their own words. (n.d.). Retrieved August 16, 2020, from https://www.advisory.com/daily-briefing/2020/06/19/mask-wearing

Wölfel, R., Corman, V. M., Guggemos, W., Seilmaier, M., Zange, S., Müller, M. A., Niemeyer, D., Jones, T. C., Vollmar, P., Rothe, C., Hoelscher, M., Bleicker, T., Brunink, S., Schneider, J., Ehmann, R., Zwirglmaier, K., Drosten, C., & Wendtner, C. (2020). Virological assessment of hospitalized patients with COVID-2019. Nature, 581, 465-469. https://doi.org/10.1038/s41586-020-2196-x

World Health Organization. (2019, June 19). Influenza. Retrieved from

World Health Organization. (2020, June 19). Coronavirus disease (COVID-19) advice for the public: When and how to use masks. Retrieved from https://www.who.int/emergencies/diseases/novel-coronavirus-2019/advice-for-public/when-and-how-to-use-masks

World Health Organization. (2020, June 29). Rolling updates on coronavirus disease (COVID-19). Retrieved from https://www.who.int/emergencies/diseases/novel-coronavirus-2019/events-as-they-happen

www.ingramcontent.com/pod-product-compliance
Lightning Source LLC
Chambersburg PA
CBHW071750270326
41928CB00013B/2868